Smart Energy Systems

The book series Smart Energy Systems comprehensively explores the application of advanced theories and cutting-edge technologies in the evolving energy landscape. The series emphasizes the strategic practical application in smart grid management, renewable energy integration, and multi-source energy coordination. The series explores topics such as data analysis, system optimization, machine learning, and AI-driven control methods, highlighting their crucial role in ensuring efficient and sustainable energy system operations. Additionally, the series broadens its focus to encompass hybrid energy solutions and distributed energy management, offering a comprehensive perspective on how smart, integrated energy systems can transform future power networks.

The book series Smart Energy Systems seeks to provide researchers, engineers, and professionals with in-depth insights into the implementation of integrated energy frameworks and renewable energy technologies. Each volume focuses on distinct aspects such as advancements in smart grid technology, real-time energy optimization, and the integration of renewable into traditional energy systems. Collectively, the series acts as an indispensable resource for academia, energy industry professionals, and policymakers, providing technical depth to bridge the gap between research and practical implementation in pursuing a sustainable energy future.

Jiajia Chen · Yuanzheng Li

Renewable Power System Optimization

Springer

Jiajia Chen
Shandong University of Technology
Zibo, Shandong, China

Yuanzheng Li
School of AI and Automation
Huazhong University of Science
and Technology
Wuhan, Hubei, China

ISSN 3059-4359 ISSN 3059-4367 (electronic)
Smart Energy Systems
ISBN 978-981-97-8131-7 ISBN 978-981-97-8132-4 (eBook)
https://doi.org/10.1007/978-981-97-8132-4

This Springer imprint is published by the registered company Springer Nature Singapore Pte Ltd.
The registered company address is: 152 Beach Road, #21-01/04 Gateway East, Singapore 189721, Singapore

Preface

In the context of global initiatives toward green development and the pursuit of carbon neutrality, the renewable power system emerges as a pivotal component of the future energy grid. This transition envisions the extensive integration of variable renewable energy such as wind and photovoltaic power into the grid infrastructure. However, this increasing reliance on variable renewable energy exacerbates the volatility of net load demand, leading to significant challenges including wind and photovoltaic power curtailment. While grid upgrades and the augmentation of flexible resource allocation can mitigate these issues, they invariably result in reduced equipment utilization, thereby introducing new challenges to the economic operation of renewable power systems. Moreover, existing methods for predicting variable renewable energy output still grapple with substantial prediction uncertainties, underscoring the considerable operational risks associated with their widespread integration into the power grid. Consequently, there is a pressing need within the renewable power sector to explore and leverage flexible resources optimally to facilitate the optimal operation of the power grid.

Against this backdrop, the exploration and optimization of flexible resources within the renewable power system assume pivotal roles in addressing the volatility and uncertainty inherent in renewable energy sources. Notably, significant strides have been made in the development of optimization techniques for flexible resources across the supply, grid, demand, and energy storage domains in recent years. However, it remains a common challenge that flexible resources are often in short supply. Given the escalating proportion of renewable energy within the energy mix, the cost associated with meeting the growing demand for flexible resources may outweigh the benefits derived from accommodating renewable energy. Such dynamics directly impact the economic viability of the power grid. Hence, this book endeavors to delve into advanced techniques for optimizing renewable power systems, with the dual objectives of showcasing valuable applications and stimulating further scholarly discourse and innovation.

This book is a monograph about renewable power system optimization across nine comprehensive chapters. It commences with an introductory overview elucidating the fundamental concepts of renewable power systems, encompassing problem statement and property. Acknowledging volatility and uncertainty as fundamental elements shaping renewable power operations and applications, subsequent chapters systematically explore optimization techniques pertinent to four crucial domains: supply side, power grid, demand side, and energy storage side. Building upon these foundational discussions, the subsequent chapters delve deeper into the escalating challenges posed by the increasing volatility and uncertainty within renewable power system optimization.

Specifically, this book addresses flexible resources optimization in renewable transmission, distribution, microgrid, and energy storage systems. Chapter 1 provides an overview elucidating the fundamental concepts of renewable power systems, including the problem statement and properties. Chapters 2 and 3 focus on renewable supply optimization, covering unit commitment and multi-energy in renewable transmission system optimization. Chapters 4 and 5 delve into network reconfiguration and optimal power flow for improving flexibility in renewable distribution system optimization. Chapters 6 and 7 investigate renewable microgrid optimization, including agricultural microgrid and park microgrid. Chapters 8 and 9 explore the application and optimization of energy storage systems.

To facilitate a clear comprehension of the breadth and depth of what we have done, the following is a succinct overview of each of the 9 chapters.

Chapter 1 motivates the subject matter of this book by introducing the challenges of volatility and uncertainty within the realm of renewable power system optimization. It provides a review of flexibility resources and optimization techniques tailored to practical applications.

Chapter 2 develops a distributionally robust unit commitment with enhanced disjointed layered ambiguity set. It only needs the probabilistic information at the vertexes of each sub-uncertainty set and eliminates the need for assuming linear generation costs.

Chapter 3 introduces the concept of fuzzy semi-entropy for quantifying the downside risk associated with uncertain renewable energy on the supply side. Subsequently, a mean-semi-entropy model is devised to optimize low-carbon-oriented multi-energy transmission system.

Chapter 4 provides an iterative bi-level scheduling for stochastic distribution system optimization to bolster both economic efficiency and system stability, considering combine multi-step network reconfiguration with many-objective reduction techniques.

Chapter 5 focuses on the optimal power flow of renewable distribution system. A novel credibility theory-based fuzzy chance-constrained AC optimal power flow model is proposed to mitigate the uncertainty surrounding renewable energy generation.

Chapter 6 offers a multi-energy hub to enhance resilience of agricultural microgrid system under the uncertainties inherent in photovoltaic and solar thermal systems, as well as their coupling relationships.

Chapter 7 derives the interval normalized Bernstein polynomial to model continuous-time renewable power system. It further develops a continuous-time optimization-driven demand power defense strategy tailored to renewable park systems.

Chapter 8 presents an expected cost minimization-driven clustering method and a multiple characteristics extraction-based recognition technique. This sets the stage for proposing a novel energy storage operation strategy for renewable park systems.

Chapter 9 proposes a peer-to-peer-based mobile energy storage optimization to improve the renewable power system resilience. The model also investigates the cost allocation mechanism among users participating in energy sharing.

In conclusion, this book collectively contributes to a comprehensive understanding of the optimization challenges and strategies within renewable power systems, paving the way for innovative solutions and advancements in the field. We hope this book can inspire readers to define new problems, employ innovative methods, and yield compelling insights by leveraging extensive historical data within power systems.

Zibo, China
Wuhan, China

Jiajia Chen
Yuanzheng Li

Acknowledgments

This book made a summary of our research about renewable power system optimization achieved in recent years. These works were carried out in the School of Electrical and Electronic Engineering, Shandong University of Technology, Zibo, China, and the School of Artificial Intelligence and Automation, Ministry of Education Key Laboratory of Image Processing and Intelligence Control, Huazhong University of Science and Technology, Wuhan, China.

Many people contributed to this book in various ways. The authors are indebted to Mr. Bingyin Xu, Mr. Yanlei Zhao, Mr. Wengang Chen, Dr. Pihua Jiao, Mr. Fengwei Liu, Mr. Jiahao Duan, Mr. Guijin Yan, Mr. Yanxin Wang, Mr. Xu Cai, Mr. Liang Feng, Miss Jingyu Li, Miss Yiqi Zhang, Miss Jingyao Liu, Miss Luwen Pan from Shandong University of Technology and Mr. Yuanzheng Li from Huazhong University of Science and Technology, who have contributed materials to this book and thank for their assistance in pointing out typos and checking the whole book.

In addition, we appreciate the staff at Springer for their assistance and help in the preparation of this book.

This work is supported by National Natural Science Foundation of China (Grant 52377110, 62422308). The authors really appreciate their supports.

Jiajia Chen
Yuanzheng Li

Contents

Part I
Fundamentals and Renewable Power Supply Optimization

Chapter 1
Introduction for Renewable Power System Optimization

The purpose of this introductory chapter is to elucidate the core themes of this book, focusing on the challenges within renewable power system optimization resulting from the escalating integration of variable renewable energy sources. It also delves into the corresponding technologies and solutions pivotal for fostering green, low-carbon development. To begin, we offer a concise exploration of the evolution and challenges intrinsic to the renewable power system. Subsequently, we delineate the flexible resources available across the supply, grid, demand, and energy storage sides, crucial for mitigating the volatility inherent in variable renewable energy within the context of renewable power system flexibility optimization. Finally, we touch upon the effective strategies for addressing the uncertainties associated with variable renewable energy through adept uncertain optimization design.

1.1 Renewable Power System Flexibility Optimization

Guided by the imperatives of carbon peak and carbon neutralization targets, variable renewable energy, exemplified by wind power and photovoltaics, has undergone rapid advancement. Concurrently, the utilization of energy storage devices and electric vehicles has seen an ever-expanding scope within the power grid. Assessing the current landscape and projecting future trajectories of power grid development, it is evident that the power network is transitioning toward a novel, intricate configuration characterized by the integration of renewable energy and the widespread adoption of diverse flexible loads. Consequently, the difference between peak and off-peak net power loads will continue to expand. Figure 1.1, often referred to as the "Duck Curve," presents the net load profile of a renewable energy-integrated power system, as analyzed by the California System Independent Operator (CAISO), a key system operation agency in the United States [1].

J. Chen and Y. Li, *Renewable Power System Optimization*, Smart Energy Systems,
https://doi.org/10.1007/978-981-97-8132-4_1

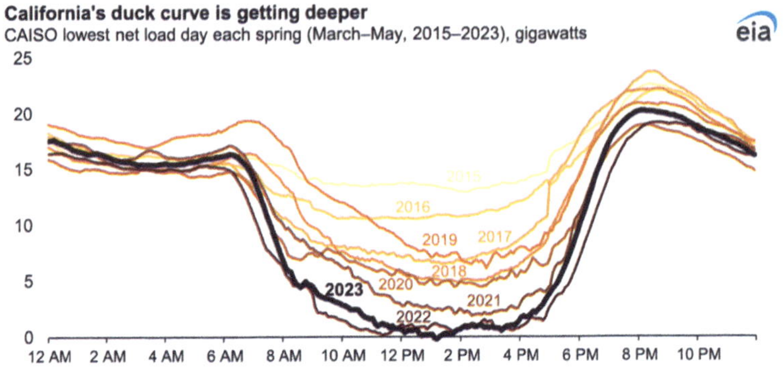

Fig. 1.1 Duck curve where california independent system operator, adopted from [1]. *Data source*: California Independent System Operator (CAISO)

The inherent variability of variable renewable energy manifests in significant random fluctuations, compounded by their substantial grid integration and temporal-spatial misalignment with various power loads. This predicament accentuates the challenge of achieving a harmonious balance between power supply and demand. Ensuring this equilibrium necessitates the rapid cycling on and off of flexible resources, primarily represented by thermal power plants, thereby diminishing equipment utilization rates. As societal electricity consumption continues its upward trajectory and renewable energy penetration deepens, the gulf between peak and off-peak net power loads will persist in widening, consequently escalating the demand for flexible resources within the power grid.

Whether adhering to the traditional scheme of infrastructure capacity expansion through distribution and transformation rotations, or embracing novel technical avenues such as low-voltage interconnections or flexible grid interlinks in recent years, the investments required are substantial while the economic returns remain modest. Consequently, the exploration of existing flexibility resources within the power grid has emerged as a focal point of interest for experts and scholars both domestically and internationally.

Flexibility optimization within this book is delineated as the capacity of a power system to effectively manage the random fluctuations inherent in variable renewable energy sources. This involves the strategic utilization of flexible resources across various time scales, encompassing both the planning and operational phases; all while adhering to physical and security constraints in a cost-effective manner. In essence, the optimization of flexibility within a renewable power system pertains to its capability to swiftly and efficiently respond to substantial, unforeseen oscillations in power supply and demand.

Within this subsection, we propose a taxonomy for the optimization of power grid flexibility, categorized across the dimensions of supply, grid, demand, and energy storage. Subsequent chapters of this book will expound upon the methodologies and models essential for delineating the potential for flexibility across all facets of renewable power system optimization.

1.1.1 Flexibility Optimization of Supply Side

Within the traditional power system framework, the optimization of flexibility predominantly relies on the generation scheduling of thermal power plants as the primary mechanism from the supply side. These plants enhance their operational flexibility through measures such as reducing start-up times, lowering minimum technical output thresholds, and improving ramping speeds to effectively accommodate shifts in load demand. However, as the proportion of renewable energy within the power system rises, traditional thermal power plants prove inadequate in meeting the dynamic flexibility demands posed by variable renewable energy sources. In recent years, three principal techniques have emerged for optimizing flexibility from the supply side: multi-source complementarity, multi-energy coupling, and the concept of a virtual power plant.

Multi-source complementarity entails the integration of various sources such as thermal power plants, hydropower units, wind power, and solar power. This approach capitalizes on the inherent complementarity among these diverse energy sources. While wind, solar, and hydropower are all renewable, they are subject to natural constraints such as wind speed, sunlight availability, and water levels. By amalgamating these energy sources, fluctuations in output can be smoothed, thereby enhancing the overall flexibility of the power system. Scholars have pursued diverse combinations of energy sources on the supply side to cater to the requirements of optimizing flexibility within renewable power systems. For instance, to mitigate the active power output fluctuations of wind power, Reference [2] advocates leveraging water storage capabilities to bolster the economic viability of wind plant operations. Reference [3] presents a coordinated optimization model for interconnected power systems encompassing hydro-thermal-wind-photovoltaic plants, aiming to minimize the curtailment of renewable energy while maximizing renewable energy generation and minimizing thermal output fluctuations. Furthermore, Reference [4] delves into the complementary operation of a hybrid pumped storage-wind-photovoltaic system across various time scales, evaluating its economic benefits and energy efficiency. These studies collectively demonstrate that multi-source complementarity not only amplifies the generation of variable renewable energy but also reduces the random output fluctuations of thermal power generation simultaneously.

Multi-energy coupling within the realm of renewable power system optimization entails the integration and orchestration of diverse energy sources, including electricity, heating, and gas, utilizing a range of energy conversion technologies. This integration forms an interconnected energy system aimed at achieving the optimal utilization of available energy sources. The primary objective is to smooth power output fluctuations and enhance the overall flexibility of the power system. For instance, Chen et al. [5] delve into the study of coordinated scheduling and optimal operational strategies for a coupled heat-power-gas system, focusing on flexibility enhancement through the implementation of cogeneration and power-to-gas technologies. Similarly, Literature [6] presents a pioneering modeling framework and associated optimization methodology tailored for short-term operational planning, with a focus on deploying multi-energy system flexibility. This framework finds application in district energy systems and facilitates participation in energy and frequency control ancillary services. The convergence of digitalization and sector coupling empowers entities to transition into "flexumers," entities capable of flexibly managing their multi-energy systems. Through the optimization of these systems, costs and carbon emissions are reduced while concurrently balancing the electricity grid [7].

Turning attention to the concept of a virtual power plant (VPP), it represents a network of decentralized, often small-scale power generation units such as solar panels, wind turbines, batteries, and demand response mechanisms. These units are amalgamated to function collectively as a single, unified power plant. Employing sophisticated software and control systems, a VPP is centrally managed to optimize the aggregate performance of these distributed resources. This centralized management facilitates better coordination and utilization of available energy, effectively aligning supply with demand [8]. The versatility of a VPP extends to providing grid stability and essential support services by regulating frequency, voltage, and overall grid equilibrium. Its capacity to swiftly respond to fluctuations in both demand and supply contributes significantly to the maintenance of a stable and reliable electricity grid [9]. In essence, a VPP epitomizes a modern and adaptable approach to energy management, leveraging digital technologies to foster an efficient, resilient, and sustainable renewable power system.

1.1.2 Flexibility Optimization of Grid Side

Grid-side flexibility optimization embodies the power grid's capacity to maintain equilibrium between generation and load amidst random fluctuations, thereby bolstering grid efficiency, resilience, and facilitating the seamless integration of variable renewables [10]. Diverging from the flexibility resources on the supply or demand sides, grid-side flexibility hinges on the regulatory capabilities, control mechanisms, and inherent parameters of the grid components. Within this domain, two quintessential techniques for flexibility optimization are Optimal Power Flow (OPF) and network reconfiguration in the context of renewable power systems.

OPF stands as a mathematical optimization approach designed to ascertain the most efficient and cost-effective manner to allocate power from diverse sources to meet the electricity demand, all while adhering to operational constraints. Tailored explicitly for renewable power systems, OPF optimizes grid operations, enabling a smoother integration of renewable energy, demand response initiatives, and energy storage systems. Previous studies have underscored OPF as a potent instrument in modernizing electricity grids. Through the utilization of advanced optimization algorithms, grid operators can strike a delicate balance between economic viability, grid reliability, and environmental sustainability. As the energy landscape evolves with the increasing integration of renewable resources and smart grid technologies, OPF assumes a pivotal role in ensuring a robust, adaptable, and efficient electricity infrastructure for the future.

On the other hand, network reconfiguration involves altering the physical layout of the power grid, such as modifying the open/closed status of switches and breakers, to optimize power flow and enhance grid performance. This method aims to amplify flexibility by dynamically adjusting the grid's configuration to align with evolving demand, generation patterns, and operational exigencies. As the energy ecosystem progresses toward decentralization and the integration of renewables, network reconfiguration emerges as a linchpin in crafting a resilient, efficient, and responsive electricity grid for the future.

1.1.3 Flexibility Optimization of Demand Side

Demand-side flexibility optimization encompasses the adoption of strategies and technologies aimed at modifying consumer behavior and electricity consumption patterns in response to grid conditions. This strategic approach seeks to shift or curtail electricity usage during peak periods, facilitate the integration of renewable energy sources, and fortify grid stability. Central to this approach are demand response programs, pivotal strategies designed to incentivize consumers to adjust their electricity consumption based on signals from the grid [11]. Such initiatives may involve rescheduling non-essential activities to off-peak hours or curbing consumption during peak periods, thereby aiding in the equilibrium of supply and demand and alleviating strain on the grid. As the energy sector transitions toward a decentralized and sustainable future, demand-side flexibility assumes an increasingly critical role in sculpting a more intelligent and efficient electricity grid.

Demand response programs are typically categorized into price-based and incentive-based mechanisms [12]. Within the realm of price-based demand response, distinctions are made based on time-of-use electricity pricing, peak electricity pricing, and real-time electricity pricing [13]. Through these pricing signals, consumers can adjust their electricity usage patterns to capitalize on periods of lower prices. This adjustment may involve running appliances, charging electric vehicles, or executing energy-intensive tasks during off-peak hours when renewable energy generation typically peaks. Moreover, price-based demand response empowers consumers to actively engage in energy markets. As the adoption of renewable energy continues to expand, price-based demand response mechanisms will play an increasingly pivotal role in enhancing the efficiency and sustainability of power systems.

Conversely, incentive-based demand response entails the formulation of corresponding policies by implementing agencies in response to the power system's supply–demand dynamics. Under this framework, consumers curtail their power demand during grid-tight or high-demand periods, in exchange for direct compensation or preferential electricity pricing during specified periods [14]. This encompasses a spectrum of strategies such as direct load control, interruptible load programs, demand-side bidding, emergency demand response, capacity market initiatives, and ancillary service projects. Through the provision of attractive rewards and benefits, utilities can stimulate consumer participation, mitigate peak demand spikes, and foster the development of a more efficient and sustainable electricity ecosystem.

1.1.4 Flexibility Optimization of Energy Storage Side

Energy storage plays a pivotal role in bridging the temporal gap between power supply and demand, thereby bestowing flexibility upon the entire system. The optimization of energy storage within the context of renewable power systems entails the utilization of diverse storage technologies to capture surplus energy generated from renewable sources during periods of heightened production. Subsequently, this stored energy can be strategically deployed during times of need, bolstering grid stability and operational efficiency [15].

Viewed from the supply side, energy storage presents an array of substantial benefits for enhancing flexibility within power systems. It enables the more efficient operation of renewable energy sources such as solar and wind power by storing excess energy when production exceeds immediate demand. This reservoir of stored energy can then be tapped into during periods of reduced renewable generation, ensuring a continuous and dependable power supply [16]. By mitigating the inherent variability of renewable sources, energy storage systems play a pivotal role in fortifying grid stability and diminishing the reliance on backup fossil fuel-based generation. This dual effect not only optimizes the utilization of renewable resources but also contributes significantly to cultivating a more efficient and sustainable energy landscape.

On the demand side, energy storage presents consumers with avenues to fine-tune their electricity consumption patterns, leading to cost savings [17]. Leveraging time-of-use pricing structures and incentives, consumers can strategically shift their electricity usage to off-peak hours when energy prices are more economical. Energy storage systems facilitate the storage of surplus energy generated by rooftop solar panels or other distributed energy resources, enabling its utilization during peak demand periods. This not only reduces electricity expenses for consumers but also empowers utilities to more effectively manage peak loads, thereby enhancing the overall efficiency and reliability of the grid. The incentivization of demand response through energy storage empowers consumers to actively engage in grid management, fostering the development of a more responsive and adaptable energy system.

From the grid perspective, energy storage assumes a pivotal role in fortifying grid flexibility and resilience. Energy storage systems offer swift-response capabilities crucial for frequency regulation, voltage support, and the provision of grid balancing services [18]. Serving as a buffer against abrupt fluctuations in supply or demand, these systems contribute significantly to grid stabilization, averting potential blackouts. Moreover, energy storage facilitates the seamless integration of emerging technologies such as electric vehicles and microgrids into the grid infrastructure. By alleviating transmission congestion, optimizing energy trading operations, and furnishing backup power during emergencies, energy storage fortifies the grid's overall capacity to accommodate variable renewable energy inputs and evolving energy demands. Consequently, it emerges as a cornerstone of modern grid infrastructure, paving the path toward a more efficient, reliable, and sustainable energy landscape for the future.

1.2 Renewable Power System Uncertainty Optimization

Presently, the predictive methodologies for renewable energy output, notably wind power and photovoltaics, still grapple with inherent limitations, failing to curtail prediction errors to a negligible range. The precision of output predictions for photovoltaic systems within the meteorological domain remains less than optimal, exhibiting forecasting errors ranging from 20 to 60% or higher during extreme weather conditions such as heavily overcast, dusty, or partly cloudy days [19]. Similarly, the prediction error for wind power hovers around 14–20% [20]. It is imperative to underscore that even with a power prediction error of 10–15%, a day-ahead energy management system struggles to attain an optimal solution [21]. Hence, uncertainty optimization strategies within renewable power systems emerge as indispensable tools for grappling with the inherent variability of renewable energy sources.

By addressing uncertainties inherent in generation forecasts, grid operators stand to enhance grid stability, reliability, and operational efficiency. Moreover, these optimization techniques facilitate optimal resource allocation, bolster risk management frameworks, and furnish decision-makers in energy markets with informed insights. As the global share of renewable energy continues its ascension, uncertainty optimization assumes an increasingly pivotal role in facilitating a seamless transition toward a sustainable and resilient energy future.

Currently, four primary models are employed for uncertain optimization within renewable power systems, namely robust optimization, fuzzy optimization, stochastic optimization, and interval optimization [22]. These models offer diverse avenues for grappling with uncertainty, each with its distinct set of advantages and applications. Through the judicious application of these uncertainty optimization models, renewable power systems can navigate the complexities of unpredictable renewable energy outputs, fostering a more reliable, efficient, and resilient energy landscape [23]. This concerted effort toward uncertainty management stands as a linchpin in shaping the trajectory of renewable energy integration and the evolution of modern energy systems toward sustainability.

1.2.1 Robust Optimization

Robust optimization stands as a cornerstone in ensuring the dependable and efficient operation of renewable power systems, which inherently grapple with uncertainties stemming from the variability of renewable energy sources [24]. Within the domain of renewable power systems, robust optimization endeavors to craft solutions that maintain efficacy across a spectrum of potential scenarios, accounting for uncertainties related to factors such as solar irradiance, wind speed, and energy demand. This strategic approach acknowledges the inherent challenges in accurately forecasting renewable energy generation and load profiles, striving to cultivate resilient solutions that exhibit robust performance under diverse conditions.

At the heart of robust optimization in renewable power systems lies the principle of minimizing the impact of uncertainty on system performance. By contemplating a set of scenarios representing varying levels of renewable energy availability, demand fluctuations, and potential disruptions, robust optimization models are adept at pinpointing solutions that exhibit resilience in the face of uncertainties. For instance, a robust optimization model designed for a solar power plant might incorporate scenarios encompassing different degrees of solar irradiance, variations in cloud cover, or equipment malfunctions. Through optimization under these diverse scenarios, the model can discern optimal operational strategies that guarantee consistent and reliable power output even amidst adverse conditions.

Robust optimization in renewable power systems offers a distinct advantage through its provision of a safety margin against uncertainties [25]. Unlike approaches solely focused on optimizing for average or expected conditions, robust optimization models account for worst-case scenarios within predefined uncertainty bounds. This proactive stance ensures the system's continued effectiveness even in scenarios where conditions deviate significantly from the norm. In the context of integrating renewable energy into the grid, robust optimization proves invaluable in tasks such as scheduling energy production, determining storage requirements, and devising contingency plans. These endeavors are undertaken with a keen awareness of the inherent variability of renewable energy sources.

The overarching impact of robust optimization is evident in its enhancement of the resilience, reliability, and efficiency of renewable power systems. By furnishing solutions robust against uncertainties, robust optimization cultivates a proactive approach to both system design and operation. Consequently, renewable energy facilities can sustain stable and predictable output levels despite the challenges posed by variable weather conditions, potential equipment failures, and fluctuating demand patterns.

1.2.2 Fuzzy Optimization

Fuzzy optimization emerges as a formidable mathematical tool within renewable power systems, adept at navigating the inherent uncertainties and imprecise information synonymous with renewable energy sources [26]. Diverging from conventional optimization methodologies reliant on precise numerical data, fuzzy optimization empowers the representation of uncertain parameters through the prism of fuzzy logic and fuzzy sets. This approach assumes paramount significance within renewable energy systems, wherein variables such as solar irradiance, wind speed, and energy demand exhibit notable variability and may lack precise quantification.

Within the realm of renewable power systems, fuzzy optimization assumes a pivotal role in facilitating decision-making amidst uncertain conditions, capturing the inherent vagueness and ambiguity of input variables [27]. For instance, forecasts pertaining to solar irradiance and wind speed often entail a degree of imprecision, leading to uncertainties regarding the resultant energy generation. Fuzzy optimization models adeptly integrate these uncertain forecasts as fuzzy sets, enabling a more nuanced depiction of the potential range of energy outputs. This inherent flexibility empowers the system to make robust decisions resilient to the vicissitudes of renewable energy production fluctuations.

An eminent advantage of fuzzy optimization within renewable power systems lies in its adeptness at handling intricate and uncertain relationships between input and output variables. Fuzzy logic rules serve as a potent tool in capturing the nonlinear and dynamic essence of renewable energy systems, duly considering factors such as weather patterns, geographical nuances, and system constraints. Through the utilization of linguistic variables and fuzzy rules, fuzzy optimization models engender decision-making processes grounded in human-like reasoning, thereby furnishing adaptive and intelligent solutions to optimize the operation of renewable power systems.

Furthermore, fuzzy optimization facilitates the seamless integration of diverse renewable energy sources—ranging from solar and wind to hydroelectric power—into a cohesive system. This integration assumes paramount importance in attaining a harmonious and efficient renewable energy mix, wherein energy production is maximized while grid stability is upheld. By optimizing the allocation and utilization of various renewable sources based on fuzzy rules and constraints, fuzzy optimization contributes substantially to the overarching reliability, efficiency, and sustainability of renewable power systems. In an era where renewable energy spearheads the transition toward a greener future, fuzzy optimization emerges as a versatile and robust framework poised to address the uncertainties inherent in renewable energy generation, thus propelling system performance optimization to new heights.

1.2.3 Stochastic Optimization

Stochastic optimization is a powerful and essential tool in the realm of renewable power systems, offering a sophisticated framework to contend with the inherent uncertainties entwined with renewable energy sources [28]. Diverging from deterministic optimization methodologies reliant on precise input data, stochastic optimization traverses the realm of randomness and variability intrinsic to factors such as solar irradiance, wind speed, and electricity demand. Within the dynamic landscape of renewable energy, where resource availability ebbs and flows unpredictably, stochastic optimization stands as a steadfast framework, enabling decisions that exude resilience and reliability amidst fluctuating conditions.

A salient strength of stochastic optimization in renewable power systems resides in its adeptness at modeling and assimilating probabilistic distributions of uncertain variables. This encompassing approach spans factors like weather patterns, which exert profound influence on the output of solar and wind farms. Harnessing historical data, weather forecasts, and an array of probabilistic inputs, stochastic optimization algorithms proficiently generate an array of scenarios depicting potential future conditions. Subsequently, these diverse scenarios serve as the bedrock for crafting optimal strategies, adept at mitigating risks and optimizing performance across a spectrum of potential outcomes.

In practical application, stochastic optimization empowers operators of renewable power systems to make judicious and informed decisions pertaining to energy production, storage, and distribution [29]. For instance, it finds utility in determining the optimal composition of renewable energy sources within a portfolio, meticulously considering the inherent variability of each source and their interrelations. Stochastic optimization further aids in the orchestration of energy generation and storage operations, deftly navigating uncertain forecasts concerning renewable energy output and electricity demand. By optimizing across a spectrum of scenarios, stochastic optimization ensures the system's preparedness for diverse contingencies, fostering dynamic adaptability to evolving conditions.

Furthermore, stochastic optimization techniques wield the potential to enhance the economic viability of renewable power systems by curtailing costs associated with uncertainty. This encompasses measures such as mitigating the necessity for costly backup power sources, sidestepping penalties attributed to under or overproduction of electricity, and fine-tuning the utilization of energy storage systems. Ultimately, stochastic optimization serves as a linchpin in augmenting the efficiency, reliability, and profitability of renewable power systems, thereby assuming a pivotal role in propelling the transition toward a sustainable and resilient energy infrastructure.

1.2.4 Interval Optimization

Interval optimization is a valuable and pragmatic approach within the domain of renewable power systems, furnishing a methodology to contend with the uncertainties and variability intrinsic to renewable energy generation [30]. This optimization paradigm diverges from traditional methods by considering parameters imbued with uncertainty as intervals, rather than fixed and precise values. Within the dynamic realm of renewable power, where variables such as solar irradiance, wind speed, and electricity demand exhibit notable fluctuations, predicting exact values proves challenging. Interval optimization rises to this challenge by delineating these variables within defined ranges or intervals, thereby fostering a conservative yet robust approach to both system design and operation.

Within the framework of interval optimization, uncertain parameters find representation as intervals encapsulating the feasible range of potential values they may assume [31]. This interval uncertainty aptly captures the inherent variability within renewable energy sources, equipment efficiencies, and demand dynamics. By diligently exploring all conceivable values within these intervals, interval optimization endeavors to unearth solutions that are not only feasible but optimal across a diverse spectrum of conditions. This prudent and conservative stance ensures that the system operates reliably, even amidst extreme scenarios or instances where precise forecasts remain elusive.

A cornerstone advantage of interval optimization within renewable power systems lies in its provision of a safety margin against uncertainties. By adopting a conservative outlook and meticulously considering worst-case scenarios within the defined intervals, interval optimization furnishes a robust solution that mitigates risks associated with the variable nature of renewable energy generation. This approach fortifies the resilience of renewable power systems, guaranteeing steadfast and dependable operation even in the face of unforeseen shifts in weather patterns or fluctuations in energy demand. Moreover, interval optimization plays a pivotal role in facilitating effective planning and management of renewable energy resources, empowering utilities and operators with the insights needed to make informed decisions while navigating the inherent uncertainties permeating the system landscape.

References

1. California System Independent Operator (CAISO). http://www.caiso.com/Documents/Briefing_DuckCurve_CurrentSystemConditions-ISOPresentation-July2023.pdf#search=duck%20curve
2. Castronuovo ED, Lopes JAP (2004) On the optimization of the daily operation of a wind-hydro power plant. IEEE Trans Power Syst 19(3):1599–1606
3. Wang X, Chang J, Meng X et al (2018) Short-term hydro-thermal-wind-photovoltaic complementary operation of interconnected power systems. Appl Energy 229:945–962

4. Wang Z, Fang G, Wen X et al (2023) Coordinated operation of conventional hydropower plants as hybrid pumped storage hydropower with wind and photovoltaic plants. Energy Convers Manage 277:116654
5. Chen JJ, Qi BX, Rong ZK et al (2021) Multi-energy coordinated microgrid scheduling with integrated demand response for flexibility improvement. Energy 217:119387
6. Corsetti E, Riaz S, Riello M et al (2021) Modelling and deploying multi-energy flexibility: the energy lattice framework. Adv Appl Energy 2:100030
7. Fleschutz M, Bohlayer M, Braun M et al (2023) From prosumer to flexumer: case study on the value of flexibility in decarbonizing the multi-energy system of a manufacturing company. Appl Energy 347:121430
8. Wang S, Wu W (2021) Aggregate flexibility of virtual power plants with temporal coupling constraints. IEEE Trans Smart Grid 12(6):5043–5051
9. Häberle V, Tayyebi A, He X et al (2024) Grid-forming and spatially distributed control design of dynamic virtual power plants. IEEE Trans Smart Grid 15(2):1761–1777
10. Li J, Liu F, Li Z et al (2018) Grid-side flexibility of power systems in integrating large-scale renewable generations: a critical review on concepts, formulations and solution approaches. Renew Sustain Energy Rev 93:272–284
11. Alrumayh O, Bhattacharya K (2019) Flexibility of residential loads for demand response provisions in smart grid. IEEE Trans Smart Grid 10(6):6284–6297
12. Paterakis NG, Erdinç O, Catalão JPS (2017) An overview of demand response: key-elements and international experience. Renew Sustain Energy Rev 69:871–891
13. Alfaverh F, Denai M, Sun Y (2023) A dynamic peer-to-peer electricity market model for a community microgrid with price-based demand response. IEEE Trans Smart Grid 14(5):3976–3991
14. Liu D, Qin Z, Hua H et al (2023) Incremental incentive mechanism design for diversified consumers in demand response. Appl Energy 329:120240
15. Garcia-Torres F, Bordons C, Tobajas J et al (2021) Stochastic optimization of microgrids with hybrid energy storage systems for grid flexibility services considering energy forecast uncertainties. IEEE Trans Power Syst 36(6):5537–5547
16. Zhang T, Ma Y, Wu Y et al (2023) Optimization configuration and application value assessment modeling of hybrid energy storage in the new power system with multi-flexible resources coupling. J Energy Storage 62:106876
17. Karimi H, Jadid S (2023) Multi-layer energy management of smart integrated-energy microgrid systems considering generation and demand-side flexibility. Appl Energy 339:120984
18. Numan M, Baig MF, Yousif M (2023) Reliability evaluation of energy storage systems combined with other grid flexibility options: a review. J Energy Storage 63:107022
19. Liu J, Fang W, Zhang X et al (2015) An improved photovoltaic power forecasting model with the assistance of aerosol index data. IEEE Trans Sustain Energy 6(2):434–442
20. Monteiro C, Bessa R, Miranda V et al (2009) Wind power forecasting: state-of-the-art 2009. Argonne national lab (ANL), Argonne, IL (United States)
21. Abdeltawab HH, Mohamed YARI (2017) Mobile energy storage scheduling and operation in active distribution systems. IEEE Trans Ind Electron 64(9):6828–6840
22. Jiao PH, Chen JJ, Peng K et al (2020) Multi-objective mean-semi-entropy model for optimal standalone micro-grid planning with uncertain renewable energy resources. Energy 191:116497
23. Deng X, Lv T (2020) Power system planning with increasing variable renewable energy: a review of optimization models. J Clean Prod 246:118962
24. Verastegui F, Lorca A, Olivares DE et al (2019) An adaptive robust optimization model for power systems planning with operational uncertainty. IEEE Trans Power Syst 34(6):4606–4616
25. Li Y, Han M, Shahidehpour M et al (2023) Data-driven distributionally robust scheduling of community integrated energy systems with uncertain renewable generations considering integrated demand response. Appl Energy 335:120749
26. Jiao PH, Chen JJ, Cai X et al (2024) Fuzzy semi-entropy based downside risk to low-carbon oriented multi-energy dispatch considering multiple dependent uncertainties. Energy 287:129717

27. Liang RH, Liao JH (2007) A fuzzy-optimization approach for generation scheduling with wind and solar energy systems. IEEE Trans Power Syst 22(4):1665–1674
28. Zakaria A, Ismail FB, Lipu MSH et al (2020) Uncertainty models for stochastic optimization in renewable energy applications. Renew Energy 145:1543–1571
29. Zhang MY, Chen JJ, Yang ZJ et al (2021) Stochastic day-ahead scheduling of irrigation system integrated agricultural microgrid with pumped storage and uncertain wind power. Energy 237:121638
30. Li Y, Wang P, Gooi HB et al (2017) Multi-objective optimal dispatch of microgrid under uncertainties via interval optimization. IEEE Trans Smart Grid 10(2):2046–2058
31. Wang B, Zhang C, Dong ZY (2020) Interval optimization based coordination of demand response and battery energy storage system considering SoC management in a microgrid. IEEE Trans Sustain Energy 11(4):2922–2931

Chapter 2
Distributionally Robust Unit Commitment with Enhanced Disjointed Layered Ambiguity Set

2.1 Introduction

The development of renewable energy (RE) has expanded significantly in response to the global energy crisis, with numerous countries targeting an 80% RE integration in their power systems by 2050 [1, 2]. This has led to the gradual formation of RE-dominated power systems [3]. However, the inherently stochastic and intermittent nature of renewable generation (RG) introduces substantial challenges to the economic and reliable operation of power systems. A particular challenge is the increased complexity of balancing load and generation in day-ahead unit commitment (UC) problems as the penetration of RE increases. Therefore, it is essential to thoroughly characterize the uncertainty associated with RGs in UC problems to mitigate their negative impacts and facilitate their integration.

Existing studies on formulating UC decisions with uncertain RGs can generally be classified into two categories: robust optimization (RO) and distributionally robust optimization (DRO). In RO, the uncertainty of RGs is addressed by formulating an uncertainty set that captures the variations in RGs [4]. By optimizing for the worst-case scenario within this uncertainty set, RO provides a reliable UC decision [5]. For example, in Ref. [6], a robust UC model is developed for scenarios with significant RE penetration by considering all potential RG outcomes and formulating the maximum RG uncertainty set. However, while the robust approach ensures no RG curtailment and load shedding under all possible RG outcomes, it tends to be overly conservative [7].

To derive more effective UC decisions, Ref. [8] utilizes historical data to formulate a narrower RG uncertainty set and proposes a new robust UC model based on this set. By incorporating historical data, this model refines the conventional maximum RG uncertainty set, leading to more practical and effective UC decisions. Nevertheless, the worst-case scenario, which the robust approach optimizes for, does not often

J. Chen and Y. Li, *Renewable Power System Optimization*, Smart Energy Systems,
https://doi.org/10.1007/978-981-97-8132-4_2

occur in reality. Moreover, without utilizing the distribution information of RGs, the decisions made by the robust approach remain conservative, particularly in cases of high RE penetration.

To address the limitations of RO methods, recent studies have applied DRO to UC problems. DRO characterizes uncertain RG using a set of probability distribution functions known as an ambiguity set [9]. The optimal UC decision is then based on the worst-case RG distribution within this set [10]. For example, in the context of power systems with significant RE penetration, day-ahead DRO-based UC models have been proposed, considering all potential RG distributions with the same mean and covariance information [11–13]. Compared to RO methods, these DRO approaches utilize the distribution information of uncertain RGs, leading to improved UC decisions.

However, several drawbacks are associated with these DRO methods. Firstly, they rely solely on the first-order moment (mean) and second-order moment (covariance) information to represent the uncertain RGs, neglecting higher-order moments that characterize the distribution of uncertain RGs more comprehensively. Secondly, the uncertain variables for RGs are assumed to be unbounded, resulting in the worst-case RG distribution including extreme and unrealistic outcomes. These drawbacks contribute to relatively conservative UC decisions. Therefore, there is a need for the development of new DRO methods that effectively handle the uncertainty of RGs in UC problems.

The incorporation of historical data in uncertainty characterization addresses some limitations of previous DRO methods. By dividing the maximum uncertainty set into sub-uncertainty sets, a new ambiguity set is created that integrates probability information from historical data within each sub-uncertainty set. These sub-uncertainty sets are layered, resulting in nested and disjoint ambiguity sets, offering a more accurate treatment of uncertainty. Based on these ambiguity sets, new DRO methods have been developed for conditioning system scheduling [14–16]. The positive outcomes of these new DRO methods inspire researchers to apply them in UC problems, leading to the development of a complex optimization model with a min–max-min structure, driving improved UC decisions [17, 18].

The application of the new DRO method to UC problems faces a significant challenge of computational complexity. The inclusion of probabilistic information within each sub-uncertainty set introduces additional constraints, thereby increasing computational demands. Moreover, using disjointed layered ambiguity sets, while enhancing the capture of uncertainty, further adds to the complexity by introducing more constraints. To address this challenge, existing studies primarily focus on adopting nested layered ambiguity sets to alleviate computational complexity. Additionally, model simplification is achieved by assuming linear generation costs [17, 18]. To further mitigate the computational burden, researchers employ the duality theorem of linear programming to reformulate the complex min–max-min structure into a more manageable min–max structure. This reformulation enables the utilization of techniques such as Benders decomposition and others [19–21]. Moreover, conventional and enhanced linear decision rules can be used to reduce computational complexity [11, 22].

Nevertheless, it is crucial to recognize that while these attempts may improve computational efficiency, they can also lead to imprecise UC decisions. Therefore, applying new DRO methods to UC problems requires developing effective reformulation approaches and solution methods to derive optimal UC decisions.

Based on the investigation of previous studies, we identify three primary research gaps:

(1) Ambiguity Sets and Extreme Outcomes: Current ambiguity sets, derived from the maximum uncertainty set, may include extreme outcomes, such as severe fluctuations in RGs. Given the substantial influence of these severe fluctuations on UC decisions [23, 24], it is crucial to develop a more precise ambiguity set that can effectively capture the uncertain RGs.
(2) Assumption of Linear Generation Costs: Existing studies that apply new DRO methods to UC problems rely on the assumption of linear generation costs to reduce model complexity. Based on the duality theorem of linear programming, the complex optimization model with a min–max-min structure is reformulated as a min–max structure. This formulation includes all additional constraints associated with probabilistic information in each sub-uncertainty set of the ambiguity, resulting in high computational demands and potentially imprecise UC decisions.
(3) Computational Burden of Benders Decomposition: Existing studies employ the Benders decomposition method to reduce computational burdens when applying new DRO methods to UC problems. However, this method still requires considerable computation time due to the optimization model's complexity. Therefore, there is a need to develop more effective and globally optimal solution methods.

To bridge these gaps, the following contributions are identified based on our work:

(1) Enhanced Disjoint Layered Ambiguity Set: To precisely capture the uncertainty of RGs, we propose an enhanced disjoint layered ambiguity set. This proposed ambiguity set not only extracts probability information from historical data but also effectively excludes extreme outcomes (severe fluctuations of RGs) by effectively utilizing this data.
(2) Two-Stage Distributionally Robust Unit Commitment (TDRUC) Model: Based on the proposed ambiguity set, we develop a TDRUC model. Through the analysis of the TDRUC model, we devise a novel reformulation approach that eliminates the need for assuming linear generation costs. This novel reformulation approach considers only the constraints related to the probabilistic information at the vertices of each sub-uncertainty set, rather than all constraints related to sub-uncertainty sets. As a result, this approach reduces model complexity, improves computational efficiency, and enables more accurate and optimal UC decisions.
(3) Mixed-Integer Quadratic Programming (MIQP) Reformulation: Based on the assumption of quadratic generation costs, the TDRUC model is reformulated as a MIQP model, which presents computational challenges. To further enhance computational efficiency in the MIQP formulation, we propose a

method called worst-case support set identification. This method is based on a proof demonstrating that the identification of the worst-case distribution can be achieved through linear programming. Specifically, we show that formulating the worst-case distribution only necessitates a small subset of vertices in the sub-uncertainty set. By reducing model complexity through the reduction of slack constraints, our method substantially decreases computation time while maintaining its applicability.

The rest of this chapter is organized as follows. An enhanced disjoint layered ambiguity set is designed in Sect. 2.2. Section 2.3 formulates the TDRUC model based on the designed ambiguity set. In Sect. 2.4, the complex TDRUC model is transformed into a novel MIQP formulation. Section 2.5 develops the worst-case support set identification method to effectively solve the MIQP. Extensive simulations are carried out in Sect. 2.6 to demonstrate the effectiveness of the designed enhanced disjoint layered ambiguity set, the novel reformulation approach, and the proposed solution method. Finally, Sect. 2.7 draws conclusions.

2.2 Enhanced Disjoint Layered Ambiguity Set

In DRO methods, uncertainty is characterized through ambiguity sets. The performance of DRO-based UC relies heavily on the capacity of ambiguity sets to accurately characterize uncertain RGs. Existing studies on DRO-based UC primarily focus on nested ambiguity sets, which are formulated based on the maximum uncertainty set and include extreme outcomes (severe fluctuations of RGs). Consequently, their worst-case distribution may involve drastic fluctuations and unrealistic deviation events of RGs, leading to relatively conservative UC decisions. Moreover, DRO-based UC problems are challenging to solve due to their complexity.

To address these issues, we propose a new ambiguity set that is flexible, easy to specify, and practical while aligning with the computational efficiency required for exact solution methods. This set, defined as an enhanced disjoint layered ambiguity set, is formulated to model renewable generation forecast error (RGFE) uncertainty. The enhanced ambiguity set leverages the probability information from historical data within disjoint layered sub-uncertainty sets. Additionally, it excludes extreme outcomes (severe fluctuations of RGs) from the maximum uncertainty set using this historical data. The proposed ambiguity set $\mathbb{F}^t$ is established as follows:

$$\mathbb{F}^t = \left\{ \mathbb{P}^t \in \mathfrak{M}(\mathcal{U}^t) \middle| \begin{array}{l} \mathbb{E}_{\mathbb{P}^t}\left\{\tilde{\mathbf{e}}_{\mathrm{RE}}^t\right\} = \mathbf{u}^t \\ \mathbb{P}^t\{\tilde{\mathbf{e}}_{\mathrm{RE}}^t \in \mathcal{U}^t\} = 1 \\ \mathbb{P}^t\{\tilde{\mathbf{e}}_{\mathrm{RE}}^t \in \mathcal{D}_k^t\} = p_k^t \\ \mathcal{U}^t = \{\tilde{\mathbf{e}}_{\mathrm{RE}}^t \in R^M \mid \\ \mathbf{e}_{\mathrm{RE}}^{t,\min} \leq \tilde{\mathbf{e}}_{\mathrm{RE}}^t \leq \mathbf{e}_{\mathrm{RE}}^{t,\max}\} \\ \mathcal{U}_k^t = \{\tilde{\mathbf{e}}_{\mathrm{RE}}^t \in R^M \mid \\ l_i^{t,j} \leq \tilde{e}_{\mathrm{RE},i}^t \leq u_i^{t,j} \\ i \in [1, ..., M], j \in [1, ..., N_i]\} \\ k = h\underbrace{(\ldots, j, \ldots)}_{M} \\ \mathcal{D}_k^t = \sum_{i=1}^{N_k} \alpha_{k,i}^t \xi_i^t, \xi_i^t \in \mathcal{U}_k^t \\ \sum_{i=1}^{N_k} \alpha_{k,i}^t = 1, \alpha_{k,i}^t \geq 0 \\ \sum_{i=k}^{K} \mathcal{U}_k^t = \mathcal{U}^t, \sum_{k=1}^{K} p_k^t = 1 \end{array} \right\} \tag{2.1}$$

where $\tilde{\mathbf{e}}_{\mathrm{RE}}^t$ denotes the RGFE random variable at time slot t, which is M-dimensional. $\tilde{\mathbf{e}}_{\mathrm{RE}}^t$ belongs to the maximum uncertainty set $\mathcal{U}^t$, and the probability of $\mathbf{e}_{\mathrm{RE}}^t \in \mathcal{U}^t$ being within the set is one. The set of all the probability distributions $\mathbb{P}^t$ included in $\mathcal{U}^t$ is expressed by $\mathfrak{M}(\mathcal{U}^t)$. These probability distributions $\mathbb{P}^t$ are refined with the constraint of deriving the expectation $\mathbf{u}^t$ (usually a zero vector) for $\tilde{\mathbf{e}}_{\mathrm{RE}}^t$. Moreover, these probability distributions $\mathbb{P}^t$ are further refined by historical data and enhanced sub-uncertainty sets $\mathcal{D}_k^t$. In detail, these distributions $\mathbb{P}^t$ are required that the realization of $\tilde{\mathbf{e}}_{\mathrm{RE}}^t$ occurs within each enhanced sub-uncertainty set $\mathcal{D}_k^t$ with a coverage probability p_k^t.

These enhanced sub-uncertainty sets $\mathcal{D}_k^t$ are M-dimensional and are formulated as the convex hull of the historical data contained in the corresponding sub-uncertainty set $\mathcal{U}_k^t$. In this chapter, the sub-uncertainty set $\mathcal{U}_k^t$ is formulated by dividing the max uncertainty set $\mathcal{U}^t$. Let N_i denote the number divisions of the max uncertainty set for ith RGFE. To simply the expression, an index function $h(\cdot)$ is used to map these disjoint layered sets into 1th, 2th, ..., and kth sub-uncertainty set.

The lower and upper bounds for ith RGFE in $\mathcal{U}_k^t$ are shown as follows.

$$l_i^{t,j} = e_{\mathrm{RE},i}^{t,\min} + (j-1)\frac{e_{\mathrm{RE},i}^{t,\max} - e_{\mathrm{RE},i}^{t,\min}}{N_i}, j \in [1, 2, \ldots, N_i] \tag{2.2}$$

$$u_i^{t,j} = e_{\mathrm{RE},i}^{t,\min} + (j)\frac{e_{\mathrm{RE},i}^{t,\max} - e_{\mathrm{RE},i}^{t,\min}}{N_i}, j \in [1, 2, \ldots, N_i] \tag{2.3}$$

In summary, the enhanced disjointed layered ambiguity set $\mathbb{F}^t$ is constructed by incorporating the enhanced sub-uncertainty sets $\mathcal{D}_k^t$, which utilize the recorded extreme outcome information and the probability information from the historical data. Illustrations of the enhanced ambiguity set and conventional disjoint layered ambiguity set both with $M = 2, N_i = 2, i \in [1, \ldots, M]$ are shown in Figs. 2.1 and 2.2, respectively.

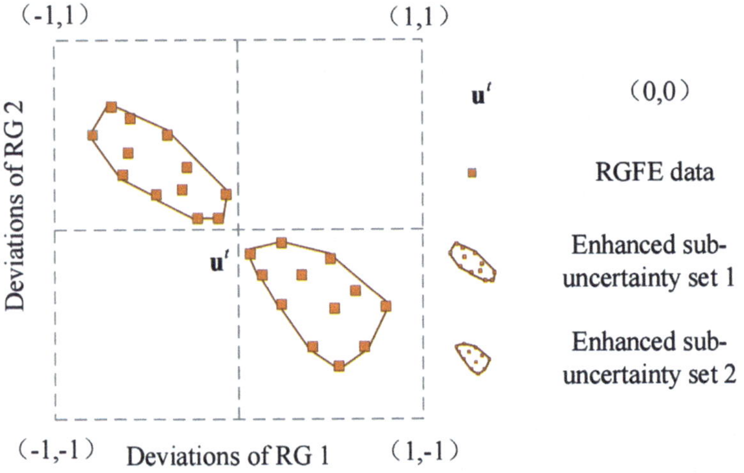

Fig. 2.1 Enhanced disjointed layered ambiguity set, reprinted from Ref. [25]. Copyright 2024, with permission IET Renewable Power Generation

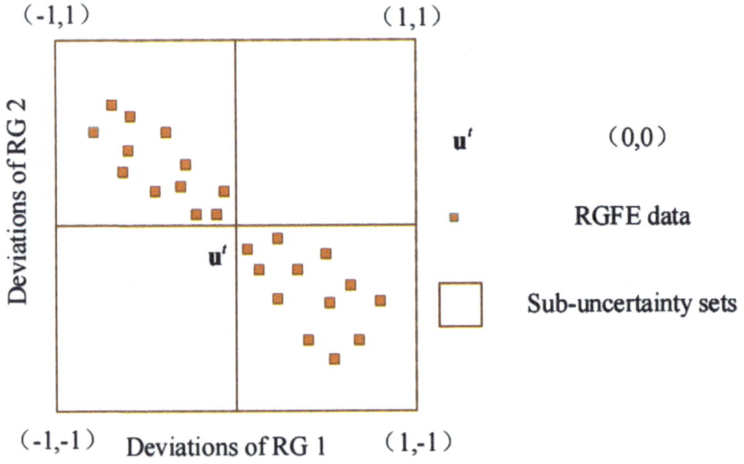

Fig. 2.2 Disjointed layered ambiguity set, reprinted from Ref. [25]. Copyright 2024, with permission IET Renewable Power Generation

In these figures, each point represents a historical RGFE data point, and each set bounded by solid lines represents the feasible space for formulating a potential distribution function. These distribution functions capture the variability and uncertainty of the RGFE data, allowing for accurate representation of the RGFE. It can be observed that the conventional disjoint layered ambiguity sets encompass wide uncertainty ranges, which may include extreme outcomes (severe fluctuations of RGs), making them less effective in capturing RGFE uncertainty. Unlike the conventional ambiguity set, the enhanced ambiguity set utilizes historical data to exclude extreme outcomes, thereby extracting more precise probability information. With the proposed ambiguity set, the DRO-based UC method can drive more effective, economic, and reliable UC decisions.

2.3 Two-Stage Distributionally Robust Unit Commitment Optimization

In this section, we formulate the TDRUC optimization model for making day-ahead economic and reliable UC decisions based on our enhanced disjoint layered ambiguity set. This optimization model employs a two-stage approach to address RGFE uncertainty. In the first stage, unit commitments are optimized before the realization of RGFEs. In the second stage, once the RGFEs are realized, the model adjusts dispatches to account for the uncertainty. The TDRUC model aims to minimize the worst-case expected total cost over the proposed ambiguity set. The details are provided in the following subsections.

2.3.1 *First-Stage Optimization Model*

$$\min_{\mathbf{x}_{\text{des}}} \sum_{t=1}^{N_T} \sum_{i=1}^{N_G} \left[(C_{\text{on},i} S_{G,\text{on},i}^t) + (C_{\text{off},i} S_{G,\text{off},i}^t) \right] \tag{2.4}$$

$$\text{s.t.}\, \mathbf{x}_{\text{des}} = (\mathbf{S}_{G,o}, \mathbf{S}_{G,\text{on}}, \mathbf{S}_{G,\text{off}}, \mathbf{P}_{G,l}, \mathbf{P}_{G,u}) \tag{2.5}$$

$$\begin{cases} \mathbf{S}_{G,o} = (S_{G,o,i}^t) \\ \mathbf{S}_{G,\text{on}} = (S_{G,\text{on},i}^t) \\ \mathbf{S}_{G,\text{off}} = (S_{G,\text{off},i}^t) \end{cases}, t \in [1, \ldots, N_T], i \in [1, \ldots, N_G] \tag{2.6}$$

$$\begin{cases} \mathbf{P}_{G,l} = (P_{G,l,i}^t) \\ \mathbf{P}_{G,u} = (P_{G,u,i}^t) \end{cases}, t \in [1, \ldots, N_T], i \in [1, \ldots, N_G] \tag{2.7}$$

$$S^t_{G,\mathrm{on},i} + S^t_{G,\mathrm{off},i} \leq 1,\ \forall t,\ i \tag{2.8}$$

$$S^t_{G,\mathrm{on},i} - S^t_{G,\mathrm{off},i} = S^t_{G,o,i} + S^{t-1}_{G,o,i},\ \forall t,\ i \tag{2.9}$$

$$S^{t-1}_{G,o,i} - S^t_{G,o,i} + S^t_{G,on,i} \geq 0,\ \forall t,\ i \tag{2.10}$$

$$S^t_{G,o,i} - S^{t-1}_{G,o,i} + S^t_{G,\mathrm{off},i} \geq 0,\ \forall t,\ i \tag{2.11}$$

$$\sum_t S^t_{G,\mathrm{on},i} \leq S^{\max}_{G,\mathrm{on},i}, \sum_t S^t_{G,\mathrm{off},i} \leq S^{\max}_{G,\mathrm{off},i},\ \forall i \tag{2.12}$$

$$S^t_{G,o,i} - S^{t-1}_{G,o,i} \leq S^\tau_{G,on,i}, \forall t, i, \tau \in [t+1,\ \min\{t + T_{u,i} - 1,\ N_T\}] \tag{2.13}$$

$$S^{t-1}_{G,o,i} - S^t_{G,o,i} \leq S^\tau_{G,\mathrm{off},i}, \forall t, i, \tau \in [t+1,\ \min\{t + T_{d,i} - 1,\ N_T\}] \tag{2.14}$$

$$P^{\min}_{G,i} S^t_{G,o,i} \leq P^t_{G,l,i} \leq P^t_{G,u,i} \leq P^{\max}_{G,i} S^t_{G,o,i},\ \forall t,\ i \tag{2.15}$$

$$P^t_{G,u,i} \leq P^{t-1}_{G,l,i} + RU_{o,i} S^{t-1}_{G,o,i} + RU_{\mathrm{on},i}(S^t_{G,o,i} - S^{t-1}_{G,o,i}) + P^{\max}_{G,i}(1 - S^t_{G,o,i}),\ \forall t,\ i \tag{2.16}$$

$$P^{t-1}_{G,u,i} \leq P^t_{G,l,i} + RD_{o,i} S^t_{G,o,i} + RD_{\mathrm{off},i}(S^{t-1}_{G,o,i} - S^t_{G,o,i}) + P^{\max}_{G,i}(1 - S^{t-1}_{G,o,i}),\ \forall t,\ i \tag{2.17}$$

where $\mathbf{x}_{\mathrm{des}}$ denotes the vector formulation of first-stage decision variables, in which $\mathbf{S}_{G,o}$, $\mathbf{S}_{G,\mathrm{on}}$, and $\mathbf{S}_{G,\mathrm{off}}$ represent the operating, start-up and shut-down states of units.

Additionally, $\mathbf{P}_{G,l}$ and $\mathbf{P}_{G,u}$ denote the lower and upper adaptive generation limits of the units [17]. Given time slots N_T and thermal units N_G, the decision variables are formulated as in Eqs. (2.6) and (2.7). Here $S^t_{G,o,i}$, $S^t_{G,\mathrm{on},i}$, $S^t_{G,\mathrm{off},i}$, $P^t_{G,l,i}$, and $P^t_{G,u,i}$ correspond to the values for unit ith at time slot tth.

In the first-stage optimization model, constraints (2.8)–(2.11) ensure the operational logic of the operating, stat-up, and shunt-down states of units to binary variables $\mathbf{S}_{G,o}$, $\mathbf{S}_{G,\mathrm{on}}$, and $\mathbf{S}_{G,\mathrm{off}}$. Constraints (2.12)–(2.14) impose limits on the number of start-up and shut-down of each unit, as well as their minimum up time and minimum shunt-down time requirements. Specifically, $S^{\max}_{G,\mathrm{on},i}$ and $S^{\max}_{G,\mathrm{off},i}$ represent the minimum start-up and the minimum shunt-down required for unit ith, while $T_{u,i}$ and $T_{d,i}$ represent the minimum up time and the minimum shunt-down time for unit ith. Additionally, first-stage constraints (2.15)–(2.17) implicitly enforce the generation capacity and ramping rate limitations [17]. In these constraints, $P^{\max}_{G,i}$ and $P^{\min}_{G,i}$ denote the upper and lower generation limits of ith unit, while $RU_{o,i}$ and $RD_{o,i}$ represent the up and down ramping limits of ith operating unit. Furthermore, $RU_{\mathrm{on},i}$ and $RD_{\mathrm{off},i}$ are the start-up and shut-down ramping limits of ith unit, respectively.

The first-stage objective, as shown in Eq. (2.4), aims to minimize the total costs associated with the start-up and shut-down of thermal units.

2.3.2 Second-Stage Optimization Model

$$\min_{\mathbf{y}_{\mathrm{des}}}\left[\sum_{t=1}^{N_T}\left[\sum_{i=1}^{N_G} C_i(P_{G,i}^t)+\sum_{i=1}^{N_B} C_B L_{B,i}^t+\sum_{i=1}^{N_{\mathrm{RE}}} C_{\mathrm{RE}} E_{\mathrm{RE},i}^t\right]\right] \tag{2.18}$$

$$\mathrm{s.t.}\mathbf{y}_{\mathrm{des}}=(\mathbf{P}_G,\mathbf{L}_B,\mathbf{E}_{\mathrm{RE}}) \tag{2.19}$$

$$\begin{cases}\mathbf{P}_G=(P_{G,i}^t),\, t\in[1,\ldots,\mathbf{N}_T],\ i\in[1,\ldots,\mathbf{N}_G]\\ \mathbf{L}_B=(L_{B,i}^t),\, t\in[1,\ldots,\mathbf{N}_T],\ i\in[1,\ldots,\mathbf{N}_B]\\ \mathbf{E}_{\mathrm{RE}}=(E_{\mathrm{RE},i}^t),\, t\in[1,\ldots,\mathbf{N}_T],\ i\in[1,\ldots,\mathbf{M}]\end{cases} \tag{2.20}$$

$$P_{G,l,i}^t\le P_{G,i}^t\le P_{G,u,i}^t,\, L_{B,i}^t\ge 0,\, E_{\mathrm{RE},i}^t\ge 0,\ \forall t,\ i \tag{2.21}$$

$$\sum_{i=1}^{N_G} P_{G,i}^t+\sum_{i=1}^{M}(\hat{P}_{\mathrm{RE},i}^t+\Delta_{\mathrm{RE},i}^t\xi_{\mathrm{RE},i}^t-E_{\mathrm{RE},i}^t)=\sum_{i=1}^{N_B}(d_i^t-L_{B,i}^t),\ \forall t \tag{2.22}$$

$$\sum_{i=1}^{N_B} F_{\mathrm{li}}\left(\sum_{j\infty i}(\hat{P}_{\mathrm{RE},j}^t+\Delta_{\mathrm{RE},j}^t\xi_{\mathrm{RE},j}^t-E_{\mathrm{RE},j}^t)+\sum_{j\infty b} P_{G,j}^t-(d_i^t-L_{B,i}^t)\right)\ge L_l^{\min},\ l\in[1,\ldots,N_L] \tag{2.23}$$

$$\sum_{i=1}^{N_B} F_{\mathrm{li}}\left(\sum_{j\infty i}(\hat{P}_{\mathrm{RE},j}^t+\Delta_{\mathrm{RE},j}^t\xi_{\mathrm{RE},j}^t-E_{\mathrm{RE},j}^t)+\sum_{j\infty b} P_{G,j}^t-(d_i^t-L_{B,i}^t)\right)\le L_l^{\max},\ l\in[1,\ldots,N_L] \tag{2.24}$$

where $\mathbf{y}_{\mathrm{des}}$ denotes the vector formulation of the second-stage decision variables, in which $\mathbf{P}_G$, $\mathbf{L}_B$, and $\mathbf{E}_{\mathrm{RE}}$ represent the generations of thermal units, the load shedding of each load node, and the power curtailment of renewable generation units. Given N_G thermal units, N_B load nodes, and M renewable generation units, decision variables are formulated as in (2.20), where $P_{G,i}^t$, $L_{B,i}^t$, and $E_{\mathrm{RE},i}^t$ are corresponding values in tth time slot for ith generation unit, ith load node, and ith renewable generation unit, respectively.

In the second-stage optimization model, constraint (2.21) restricts the dispatch of generation units $\mathbf{P}_G$ and ensures that potential load shedding $\mathbf{L}_B$ and renewable power curtailment $\mathbf{E}_{\mathrm{RE}}$ are non-negative. Constraint (2.22) enforces the power balance of the system, considering the forecasted renewable power $\hat{P}^t_{\mathrm{RE},i}$, thermal unit generations $P^t_{G,i}$, the load shedding $L^t_{B,i}$, and the renewable power curtailment $E^t_{\mathrm{RE},i}$, for the realization of RGFEs ξ_{RE}. Line security is maintained by setting line transmission capacity limits, as specified by constraints (2.23)–(2.24). In these constraints, the branch power is determined using the generation shift distribution factor F_{li}, and the lower and upper limits of line transmission capacity are represented by $L_l^{\min}$ and $L_l^{\max}$.

The second-stage objective function, as given in Eq. (2.18), aims to minimize the total costs associated with the generation of thermal units, load shedding at load nodes, and the curtailment of renewable power from generation units.

2.3.3 Compact Optimization Model Formulation for TDRUC

$$\min_{\mathbf{x}_{\mathrm{des}}} f(\mathbf{x}_{\mathrm{des}}) + \sup_{\mathbb{P}\in\mathbb{F}} \mathbb{E}_{\mathbb{P}}\left[Q(\mathbf{x}_{\mathrm{des}}, \mathbf{y}_{\mathrm{des}}, \tilde{\mathbf{e}}_{\mathrm{RE}})\right] \tag{2.25}$$

$$\text{s.t.} \mathbf{A}\mathbf{x}_{\mathrm{des}} \le \mathbf{b}, \mathbf{x}_{\mathrm{des}} = (\mathbf{S}_{G,o}, \mathbf{S}_{G,\mathrm{on}}, \mathbf{S}_{G,\mathrm{off}}, \mathbf{P}_{G,l}, \mathbf{P}_{G,u}) \tag{2.26}$$

where $Q(\mathbf{x}_{\mathrm{des}}, \mathbf{y}_{\mathrm{des}}, \boldsymbol{\xi}_{\mathrm{RE}})$ can be equivalent to

$$\min_{\mathbf{y}_{\mathrm{des}}} g(\mathbf{y}_{\mathrm{des}}) \tag{2.27}$$

$$\text{s.t.} \mathbf{B}\mathbf{x}_{\mathrm{des}} + \mathbf{C}\mathbf{y}_{\mathrm{des}} + \mathbf{D}_{\mathrm{RE}} \ge \mathbf{d}, \mathbf{y}_{\mathrm{des}} = (\mathbf{P}_G, \mathbf{L}_B, \mathbf{E}_{\mathrm{RE}}) \tag{2.28}$$

where $\mathbf{x}_{\mathrm{des}}$ represents the first-stage decision variables $(\mathbf{S}_{G,o}, \mathbf{S}_{G,\mathrm{on}}, \mathbf{S}_{G,\mathrm{off}}, \mathbf{P}_{G,l}, \mathbf{P}_{G,u})$, and $\mathbf{y}_{\mathrm{des}}$ denotes the second-stage decision variables. Additionally, constraint (2.26) provides a compact formulation of the first-stage constraints (2.8)–(2.16), while constraint (2.28) offers a compact formulation of the second-stage constraints (2.21)–(2.24).

To minimize the total cost under the worst realization of RGFE distributions, the objective function in Eq. (2.25) is defined as the sum of the first-stage start-up and shut-down costs $f(\mathbf{x}_{\mathrm{des}})$ and the second-stage worst-case expected operation cost $\sup_{\mathbb{P}\in\mathbb{F}} \mathbb{E}_{\mathbb{P}}\left[Q(\mathbf{x}_{\mathrm{des}}, \mathbf{y}_{\mathrm{des}}, \tilde{\mathbf{e}}_{\mathrm{RE}})\right]$. Here, $\tilde{\mathbf{e}}_{\mathrm{RE}}$ represents the RGFE random variable, and the potential distributions are characterized by the designed ambiguity set $\mathbb{F}$. For each realization of RGFE ξ_{RE} indicates the second-stage optimization model, with $Q(\mathbf{x}_{\mathrm{des}}, \mathbf{y}_{\mathrm{des}}, \boldsymbol{\xi}_{\mathrm{RE}})$ denoting the operation cost associated with this realization.

It is important to note that the TDRUC optimization model is formulated with a min–max-min structure, akin to a two-stage UC model based on RO. In the RO framework, the objective is to minimize the total operational cost under the worst-case realization of RGFE within an uncertainty set, leading to a three-level min–max-min optimization model. In this structure, the top-level and bottom-level models focus on minimizing UC and generation costs, respectively, while the middle-level model addresses the maximization problem to identify the worst-case uncertainty realization.

In contrast, the DRO framework aims to find the worst-case realization of the distribution function of uncertain RGFEs. This approach introduces significant computational complexity due to the need to incorporate distributional information and search for the worst-case distribution. Consequently, solving the TDRUC optimization model necessitates advanced techniques to effectively manage this added complexity.

2.4 Reformation Approach and Worst-Case Support Set Identification

To address the complexity inherent in the three-level min–max-min structure of the TDRUC optimization model, existing studies typically focus on reformulating the model into a more manageable min–max structure and utilizing the Benders decomposition method for solution. Although Benders decomposition is effective for models with a min–max structure, the incorporation of distributional information and the search for the worst-case distribution in the TDRUC model introduce additional complexities, significantly reducing the method's effectiveness.

To effectively solve the complex TDRUC optimization model, this section offers a comprehensive analysis and proposes a novel reformulation. The analysis begins by noting that the worst-case expectation in the second-stage optimization model $\sup_{\mathbb{P}\in\mathbb{F}} \mathbb{E}_{\mathbb{P}}(Q(\mathbf{x}_{\text{des}}, \mathbf{y}_{\text{des}}, \tilde{\mathbf{e}}_{\text{RE}}))$ can be decomposed into N_T independent sub-problems. Each sub-problem, denoted as $\sup \mathbb{E}_{\mathbb{P}^t}\left(Q(\mathbf{x}^t_{\text{des}}, \mathbf{y}^t_{\text{des}}, \tilde{\mathbf{e}}_{\text{RE}})\right)$, is then converted into its dual formulation with complex constraints in a max–min structure. Subsequently, under suitable assumptions, these complex constraints are reformulated into more tractable forms. Based on this analysis, a novel reformulation of the TDRUC model is developed. Detailed descriptions of this reformulation are provided in the following subsections.

2.4.1 Equivalent Separation of the Worst-Case Expectation

The worst-case expectation $\sup_{\mathbb{P}\in\mathbb{F}} \mathbb{E}_{\mathbb{P}}(Q(\mathbf{x}_{\text{des}}, \mathbf{y}_{\text{des}}, \tilde{\mathbf{e}}_{\text{RE}}))$ depends on both the second-stage optimization model $Q(\mathbf{x}_{\text{des}}, \mathbf{y}_{\text{des}}, \xi_{\text{RE}})$ and the enhanced ambiguity set $\mathbb{F}$. The designed ambiguity set is the Cartesian product of ambiguity sets $\mathbb{F}^t$ for all time slots, indicating that each ambiguity set is independent. Furthermore, in the TDRUC model, the first-stage constraints (2.12)–(2.14) implicitly enforce intertemporal constraints on generation and ramping capacity. Consequently, the dispatch optimization model in the second stage for each time slot becomes independent. Thus, the worst-case expected operation cost $\sup_{\mathbb{P}\in\mathbb{F}} \mathbb{E}_{\mathbb{P}}(Q(\mathbf{x}_{\text{des}}, \mathbf{y}_{\text{des}}, \tilde{\mathbf{e}}_{\text{RE}}))$ is equivalent to solving N_T independent sub-problems, as detailed below.

$$\sup_{\mathbb{P}\in\mathbb{F}} \mathbb{E}_{\mathbb{P}}(Q(\mathbf{x}_{\text{des}}, \mathbf{y}_{\text{des}}, \tilde{\mathbf{e}}_{\text{RE}})) = \sum_t \sup_{\mathbb{P}^t\in\mathbb{F}^t} \mathbb{E}_{\mathbb{P}^t}\left(Q^t(\mathbf{x}_{\text{des}}, \mathbf{y}^t_{\text{des}}, \tilde{\mathbf{e}}^t_{\text{RE}})\right)$$

2.4.2 Reformulation of the Worst-Case Expectation

Based on the separation of the worst-case expectation, concentrate on its reformulation at time slot tth, i.e., $\sup_{\mathbb{P}^t\in\mathbb{F}^t} \mathbb{E}_{\mathbb{P}^t}\left(Q^t(\mathbf{x}_{\text{des}}, \mathbf{y}^t_{\text{des}}, \tilde{\mathbf{e}}^t_{\text{RE}})\right)$. According to the definition of the proposed ambiguity set $\mathbb{F}^t$ in (2.1), the worst-case expectation can be written as a semi-infinite optimization model (2.29)–(2.31) of the probability measure $\mathbb{P}^t$ as follows.

$$\sup_{\mathbb{P}^t} \int_{\mathcal{U}^t} Q(\mathbf{x}^t_{\text{des}}, \mathbf{y}^t_{\text{des}}, \xi^t_{\text{RE}}) d\mathbb{P}^t(\xi^t_{\text{RE}}) \tag{2.29}$$

$$\text{s.t.} \int_{\mathcal{U}^t} \xi^t_{\text{RE}} d\mathbb{P}^t(\xi^t_{\text{RE}}) = \mathbf{u}^t \tag{2.30}$$

$$\int_{\mathcal{D}^t_k} d\mathbb{P}^t(\backslash\xi^t_{\text{RE}}) = p^t_k, \ \forall k \tag{2.31}$$

The objective function (2.29) in the aforementioned model can be further reformulated as follows:

$$\begin{aligned} &\sup_{\mathbb{P}^t} \int_{\mathcal{U}^t} Q(\mathbf{x}^t_{\text{des}}, \mathbf{y}^t_{\text{des}}, \xi^t_{\text{RE}}) d\mathbb{P}^t(\xi^t_{\text{RE}}) \\ &= \sum_{i=k}^{K} \int_{\mathcal{U}^t} Q(\mathbf{x}^t_{\text{des}}, \mathbf{y}^t_{\text{des}}, \xi^t_{\text{RE}}) d\mathbb{P}^t(\xi^t_{\text{RE}}) \\ &= \sum_{i=k}^{K} \int_{\mathcal{D}^t_k} Q(\mathbf{x}^t_{\text{des}}, \mathbf{y}^t_{\text{des}}, \xi^t_{\text{RE}}) d\mathbb{P}^t(\xi^t_{\text{RE}})\% \end{aligned} \tag{2.32}$$

The first equation holds because, in the enhanced disjoint layered ambiguity set, the maximum uncertainty set $\mathcal{U}^t$ is divided into K sub-uncertainty sets $\mathcal{U}^t_k$. The union of these sub-uncertainty sets covers the entire uncertainty set, i.e., $\sum_{i=k}^{K} \mathcal{U}^t_k = \mathcal{U}^t$. This implies that the objective function can be equivalently expressed as the sum of the objectives within each sub-uncertainty set. The second equation holds because the sub-uncertainty sets $\mathcal{U}_k^t$ are further refined into $\mathcal{D}_k^t$ by effectively utilizing historical data. By refining $\mathcal{U}_k^t$, the proposed ambiguity set considers only the distribution support within the enhanced sub-uncertainty sets $\mathcal{D}_k^t$, i.e., $\mathbb{P}(x) = 0$ if $x \notin \mathcal{D}_k^t$. Consequently, the optimization is performed within a more refined and realistic distribution support, improving the efficiency of the TDRUC model.

Based on this, the semi-infinite optimization model is transformed into the following model (2.33)–(2.34) by taking its dual.

$$\inf_{\pi^t, v^t} (\pi^t)^T \mu^t + \sum_k v_k^t p_k^t \tag{2.33}$$

$$\text{s.t.} Q^t(\mathbf{x}, \mathbf{y}^t, \xi_{\text{RE}}^t) - (\pi^t)^T \xi_{\text{RE}}^t - v_k^t \leq 0, \forall \xi_{\text{RE}}^t \in \mathcal{D}_k^t, k \in [1, \ldots, K] \tag{2.34}$$

where π^t in R^M and v^t in R^K are introduced dual variables. Constraint (2.34) are rather complex. First, they are infinite inequality constraints, i.e., for each RGFE data ξ_{RE}^t belongs to the enhanced sub-uncertainty set $\mathcal{D}_k^t$, the inequality constraint should hold. Second, $Q^t(\mathbf{x}, \mathbf{y}^t, \xi_{\text{RE}}^t)$ in constraint (2.34) indicates the second-stage optimization model, which means they can be more precisely expressed as follows.

$$\max_{\xi_{\text{RE}}^t \in D_k^t} \min_{\mathbf{y}^t \in \mathcal{Y}(\mathbf{x}, \xi_{\text{RE}}^t)} g^t(\mathbf{y}^t) - (\pi^t)^T \xi_{\text{RE}}^t - v_k^t \leq 0, \forall k \in [1, \ldots, K] \tag{2.35}$$

2.4.3 Novel Reformulation of Complex Constraints

The complex constraints with the max–min structure presented above are challenging to solve. Existing studies predominantly focus on reformulating the min–max structure into a more tractable form using the duality theorem of linear programming. In contrast to these traditional approaches, we propose a novel reformulation for these complex constraints.

It is noteworthy that, for any given $\mathbf{x}$, $Q^t(\mathbf{x}, \mathbf{y}^t, \xi_{\text{RE}}^t)$ can be viewed as a function of ξ_{RE}^t. Remarkably, we observe that this function is convex with respect to ξ_{RE}^t under suitable assumptions, as demonstrated by the following theorem.

Theorem 1 *Under the following assumptions:*

1. Line transmission limits are omitted.

2. $C_i(P_{G,i}^t)$ *is convex with respect* $P_{G,i}^t$.

3. $C_B \geq \max\left(\frac{dC_i(P_{G,i}^t)}{dP_{G,i}^t}\right), \forall i.$

4. $C_{\text{RE}} \geq 0$.

Then, for any given first-stage decision $\mathbf{x}_{\text{des}}$, $Q^t(\mathbf{x}, \mathbf{y}^t, \xi_{\text{RE}}^t)$ *is convex with* ξ_{RE}^t.

Theorem 1 indicates that the complex constraints can be reformulated as follows:

$$\max_{\xi_{\mathrm{RE}}^t \in \mathcal{D}_k^t} Q^t(\mathbf{x}, \mathbf{y}^t, \xi_{\mathrm{RE}}^t) - (\pi^t)^T \xi_{\mathrm{RE}}^t - v_k^t \leq 0, \ \forall k \in [1, \ldots, K] \tag{2.36}$$

In above formulation, $(\pi^t)^T \xi_{\mathrm{RE}}^t$ is a liner function of ξ_{RE}^t. Thus, $Q^t(\mathbf{x}, \mathbf{y}^t, \xi_{\mathrm{RE}}^t) - (\pi^t)^T \xi_{\mathrm{RE}}^t$ is also convex with ξ_{RE}^t. Moreover, $\mathcal{D}_k^t$ is a convex set. Since the maximum of a convex function over a convex set is achieved at the vertices [26], the above constraint can be reformulated as follows:

$$\bigcup_i \zeta_{k,i}^t = \mathrm{vertex}\{\mathcal{D}_k^t\}, k \in [1, \ \ldots, \ K], i \in [1, \ \ldots, \ N_{\mathrm{v},k}] \tag{2.37}$$

$$Q^t(\mathbf{x}_{\mathrm{des}}, \mathbf{y}_{\mathrm{des}}^t, \zeta_{k,i}^t) - (\pi^t)^T \zeta_{k,i}^t - v_k^t \leq 0, \ \forall k, \ i \tag{2.38}$$

where $\zeta_{k,i}^t$ denotes ith vertex in the enhanced sub-uncertainty set $\mathcal{D}_k^t$. Constraint (2.37) ensures all vertexes in the set are considered. In this constraint, $N_{\mathrm{v},k}$ represents the number of vertexes in kth set.

2.4.4 Novel Reformulation of TDRUC Optimization Model

Based on the above analysis, the TDRUC optimization model (2.25)–(2.28) is reformulated into the following novel formulation, incorporating the auxiliary variables $\theta_{k,i}^t$:

$$\min_{\mathbf{x}_{\mathrm{des}}, \mathbf{y}_{\mathrm{des}}, \pi, v, \theta} f(\mathbf{x}_{\mathrm{des}}) + \sum_t (\pi^t)^T \mu^t + \sum_k v_k^t p_k^t \tag{2.39}$$

$$\text{s.t. } \mathbf{A}\mathbf{x}_{\mathrm{des}} \leq \mathbf{b}, \mathbf{x}_{\mathrm{des}} = (\mathbf{S}_{G,o}, \mathbf{S}_{G,\mathrm{on}}, \mathbf{S}_{G,\mathrm{off}}, \mathbf{P}_{G,l}, \mathbf{P}_{G,u}) \tag{2.40}$$

$$\bigcup_i \zeta_{k,i}^t = \mathrm{vertex}\{\mathcal{D}_k^t\}, \ \forall k, \ i, \ t \tag{2.41}$$

$$g(\mathbf{y}_{\mathrm{des},k,i}^t) \leq \theta_{k,i}^t, \ \forall k, \ i, \ t \tag{2.42}$$

$$\theta_{k,i}^t - (\pi^t)^T \zeta_{k,i}^t - v_k^t \leq 0, \ \forall k, \ i, \ t \tag{2.43}$$

$$\mathbf{B}\mathbf{x}_{\mathrm{des}} + \mathbf{C}\mathbf{y}_{\mathrm{des},k,i}^t + \mathbf{D}\zeta_{k,i}^t \geq \mathbf{d}, \ \forall k, \ i, \ t \tag{2.44}$$

$$\mathbf{y}_{\mathrm{des},k,i}^t = (\mathbf{P}_{G,k,i}^t, \mathbf{L}_{B,k,i}^t, \mathbf{E}_{\mathrm{RE},k,i}^t), \ \forall k, \ i, \ t \tag{2.45}$$

Note that $g^t(\mathbf{y}^t)$ represents the objective function of the second-stage optimization model. When linear generation costs are assumed, these constraints are linear. However, with quadratic generation costs, these constraints become quadratic. Consequently, the MILP or MIQP model defined by (2.39)–(2.45) can now be addressed using most existing solvers, such as Gurobi.

2.5 Worst-Case Support Set Identification Method

Although the optimization model (2.39)–(2.45) can be directly addressed by existing commercial solvers, this approach often leads to high computational costs, particularly under the quadratic generation cost assumption. The MIQP model becomes complex due to the quadratic constraints and the large scale of the model. Specifically, the number of constraints (2.42)–(2.45) depends on the number of vertices in the enhanced sets $\mathcal{D}$. If all vertices are considered, $N_{\mathrm{T}} \sum_{i=1}^{K} N_{\mathrm{v},k}$ constraints will be introduced, where $N_{\mathrm{v},k}$ represents the number of vertices in the enhanced set.

To effectively tackle this large-scale optimization problem, we propose a worst-case support set identification method. This method is based on the observation that only a small subset of vertices is crucial for identifying the worst-case distribution. The remaining vertices and their associated constraints can be relaxed without affecting the optimal objective value. This approach reduces the model's scale and significantly enhances computational efficiency. The detailed procedure for the proposed solution method is outlined in the following sections.

Step 1: Formulate the optimization model (2.39)–(2.45) and then relax constraints (2.41)–(2.45).

Step 2: Identify the worst-case distribution support set for the second-stage optimization model, excluding line transmission limits (2.23)–(2.24), based on Theorem 2.

Theorem 2 *Under the same assumptions as Theorem 1, for any given first-stage decision* $\mathbf{x}_{\mathrm{des}}$*, the worst-case distribution support set for the second-stage dispatch at time slot* tth *can be determined using the vertices* $\zeta_{k,i}^t$*. These vertices* $\zeta_{k,i}^t$ *are part of the optimal solution to the following optimization problem with* $p_{k,i}^t > 0$*:*

$$\max_{p_{k,i}^t} \sum_{k,i} \left| \tau^t + \sum_{r=1}^{M} \zeta_{k,i,r}^t \Delta_{\mathrm{RE},r}^t \right| p_{k,i}^t \tag{2.46}$$

$$\text{s.t.} \left| \tau^t + \sum_{r=1}^{M} u_r^t \Delta_{\mathrm{RE},r}^t \right| = 0 \tag{2.47}$$

$$\sum_{k,i} \zeta_{k,i}^t p_{k,i}^t = u^t, \ \sum_{i} p_{k,i}^t = p_k^t, \ p_{k,i}^t \geq 0 \tag{2.48}$$

Step 3: Incorporate the identified support set (i.e., a subset of vertexes) into the model (2.39)–(2.45) by adding the corresponding constraints (2.42)–(2.45). This results in a relaxed formulation of the optimization model. Solve this relaxed model to obtain the optimal solution $(\mathbf{x}_{\mathrm{des}}^*, \mathbf{y}_{\mathrm{des}}^*)$.
Step 4: Verify the optimality of the solution by checking whether the identified worst-case distribution support set is a subset of the valid worst-case distribution support set for the original optimization model (2.39)–(2.45). If this condition is met, the proposed method terminates with a globally optimal solution. If not, proceed to identify additional vertexes based on the optimization model (2.39)–(2.45) with the fixed first-stage decision $\mathbf{x}_{\mathrm{des}}^*$.
Step 5: Add the newly identified support set to the previous relaxed formulation of the optimization model (2.39)–(2.45) by updating constraints (2.42)–(2.45). Solve the new relaxed model to obtain the updated optimal solution $(\mathbf{x}_{\mathrm{des}}^*, \mathbf{y}_{\mathrm{des}}^*)$.
Step 6: Repeat Steps 4 and 5 until convergence.
The flowchart of the proposed worst-case support set identification method is illustrated in Fig. 2.3. It shows that the optimization model expands in each iteration as constraints from previous iterations are retained. In the worst case, all constraints related to the vertexes in the enhanced sets $\mathcal{D}_k$ will be included in the relaxed model. Thus, the result will be globally optimal, and the iteration process will terminate. Therefore, the proposed method is guaranteed to converge.

2.6 Simulation Analysis

This section presents simulation results to validate the effectiveness of the proposed ambiguity set, the novel reformulation approach, and the solution method for the TDRUC model.

2.6.1 Simulation Settings

Simulations were conducted using a modified IEEE 24-bus power system. This system integrates a wind farm and a photovoltaic (PV) station, each with a capacity of 35 MW, connected to buses 2 and 22, respectively. The forecasted wind and PV power generation data are sourced from [27]. To simulate historical forecast errors and construct the ambiguity set, 3000 random realizations were sampled from a multivariate normal distribution with a mean of 0. The standard deviations were set at 20 and 13% of the forecasted wind and PV generation, respectively, with a correlation coefficient of -0.35 [28]. Additionally, risk factors for renewable generation curtailment and load shedding were set at \$300/MWh and \$3000/MWh, respectively [29]. All simulations were performed in MATLAB on an Intel Core-i5 3.3 GHz laptop with 16 GB of RAM, using YALMIP for modeling and Gurobi as the solver [30, 31].

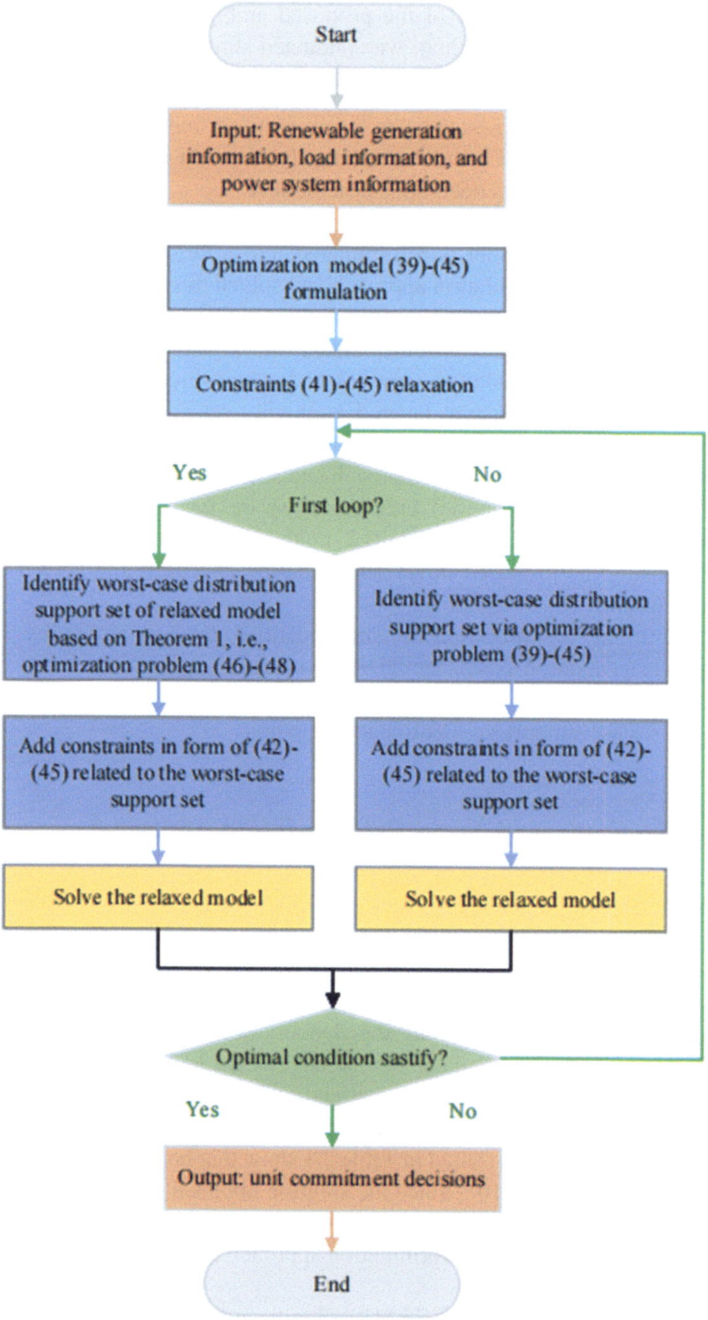

Fig. 2.3 Flowchart of the worst-case support set identification method, reprinted from Ref. [25]. Copyright 2024, with permission IET Renewable Power Generation

To evaluate the performance of the proposed ambiguity set, novel reformulation approach, and solution method, we conducted simulations comparing various scenarios. These scenarios include:

(1) M1: Using the existing nested layered ambiguity set and assuming linear generation costs, the TDRUC model is reformulated into a min–max structure and solved with the Benders decomposition method.
(2) M2: Utilizing the existing nested layered ambiguity set and linear generation costs, the TDRUC model is reformulated into a large-scale MILP with the proposed novel reformulation approach and solved directly using Gurobi.
(3) M3: With the existing nested layered ambiguity set and quadratic generation costs, the TDRUC model is reformulated into a large-scale MIQP using the proposed novel reformulation approach and solved directly using Gurobi.
(4) M4: Employing the disjoint layered ambiguity set and quadratic generation costs, the TDRUC model is reformulated into a large-scale MIQP with the proposed novel reformulation approach and solved directly using Gurobi.
(5) M5: Applying the proposed enhanced disjoint layered ambiguity set and quadratic generation costs, the TDRUC model is reformulated into a large-scale MIQP using the proposed novel reformulation approach and solved directly using Gurobi.
(6) M6: Utilizing the proposed enhanced disjoint layered ambiguity set and quadratic generation costs, the TDRUC model is reformulated into a large-scale MIQP with the proposed novel reformulation approach and solved using the worst-case support set identification method.

2.6.2 Effectiveness of Novel Reformulation Approach

In this subsection, simulations are conducted to validate the effectiveness of the proposed novel reformulation approach by comparing two scenarios: M1 and M2. Both scenarios assume linear generation costs and employ the same existing nested layered ambiguity set. Specifically, in M1, the TDRUC model is reformulated into a min–max structure and solved using the Benders decomposition method. In M2, the TDRUC model is reformulated into a large-scale MILP using our proposed novel formulation approach and solved directly using Gurobi. The results of these simulations are summarized in the following table.

The numerical results presented in Table 2.1 show that both M1 and M2 yield the same total cost of \$$6.20 \times 10^4$. Since the solution obtained by M1, which uses the Benders decomposition method, is globally optimal, the identical total costs for M1 and M2 confirm that the novel reformulation approach used in M2 also achieves the optimal solution.

However, the computation time for M1 is significantly longer than for M2. Specifically, M1 requires 1.76×10^3 s, while M2 completes in only 6.65×10^1 s. This disparity in computation time highlights the effectiveness of the proposed novel reformulation approach employed in M2.

Table 2.1 Comparison of M1-M2, reprinted from Ref. [25]. Copyright 2024, with permission IET Renewable Power Generation

	Total cost ($)	Computation time (s)	Iteration number
M1	6.20×10^4	1.76×10^3	6
M2	6.20×10^4	6.65×10^1	1

The novel reformulation approach effectively simplifies the complex constraints associated with probabilistic information at sub-uncertainty sets, as described in constraints (2.35). Unlike the conventional approach, which considers all conditions within the sub-uncertainty set, the proposed method, based on Theorem 1, reduces model complexity by focusing only on the vertexes within each sub-uncertainty set. By concentrating on the constraints related to the probabilistic information at these vertexes, the novel reformulation approach minimizes model complexity and enhances computational efficiency while maintaining solution optimality.

2.6.3 *Effect of Quadratic Generation Cost Assumption*

In this subsection, we compare the simulation results of M2 and M3 to assess the impact of assuming quadratic generation costs. Both M2 and M3 use the existing nested layered ambiguity set and the proposed novel reformulation approach. M2 reformulates the TDRUC model into a large-scale MILP under the assumption of linear generation costs, while M3 reformulates it into a large-scale MIQP with quadratic generation costs. The detailed simulation results are presented below.

The simulation results summarized in Table 2.2 reveal that M2 results in a higher total cost of $\$6.20 \times 10^4$ compared to M3, which has a total cost of $\$5.60 \times 10^4$. The elevated cost for M2 is attributed to the linear generation cost assumption, which tends to overestimate costs and consequently leads to the use of less efficient generation units. This overestimation results in higher start-up, shut-down, generation, and risk costs for M2, as detailed in Table 2.3. Consequently, the total cost for M2 exceeds that of M3 by more than 10.7%. This suggests that the quadratic generation cost assumption used in M3 provides more accurate and effective UC decisions than the linear assumption in M2.

Table 2.2 Comparison of M2-M3, reprinted from Ref. [25]. Copyright 2024, with permission IET Renewable Power Generation

	Total cost ($)	Computation time (s)	Iteration number
M2	6.20×10^4	6.65×10^1	1
M3	5.60×10^4	3.60×10^3	1

Table 2.3 Comparison of optimal solutions in M1-M2, reprinted from Ref. [25]. Copyright 2024, with permission IET Renewable Power Generation

	Start up and shut-down cost ($)	Generation cost ($)	Risk cost ($)
M2	1.70×10^3	5.58×10^4	4.50×10^3
M3	1.60×10^3	5.00×10^4	4.31×10^3

However, it is important to note that the computation time for M3 is 3.60×10^3 significantly longer than for M2, with M3 requiring more than s compared to M2's 6.65×10^1 s. This increase in computation time is due to the greater complexity of the MIQP model, which includes quadratic constraints, compared to the linear constraints of the MILP model. In summary, while the quadratic generation cost assumption offers more precise and optimal UC decisions, it also introduces a greater computational burden in solving the TDRUC optimization model.

2.6.4 Effectiveness of Designed Ambiguity Set

In this subsection, simulations are conducted to assess the effectiveness of the proposed enhanced disjoint layered ambiguity set using M3, M4, and M5. Here, M3 employs the existing nested layered ambiguity set, M4 utilizes the disjoint layered ambiguity set, and M5 is based on the newly designed ambiguity set. For all ambiguity sets, the number of sub-uncertainty sets is fixed at 12. Specifically, in M3, the parameter α\alphaα is varied from 0.75 to 2 to control the size of the nested layered sub-uncertainty sets. The resulting simulation results for these parameter settings are summarized in the following table.

Based on the results presented in Table 2.4, it is evident that M3, which employs the existing nested layered ambiguity set, incurs the highest total and generation costs among all methods. This suggests that the existing nested layered ambiguity set tends to overestimate the impact of RGFE, leading to sub-optimal UC decisions. In contrast, M4, which utilizes the disjoint layered ambiguity set, achieves lower total and generation costs compared to M3, highlighting the improved effectiveness of the disjoint layered ambiguity set in capturing RGFE uncertainty. Furthermore, M5, which is based on the designed enhanced disjoint layered ambiguity set, attains the lowest total, generation, and risk costs. This indicates that the designed ambiguity set (M5) more accurately characterizes RGFE uncertainty, particularly in high-risk scenarios. Therefore, the TDRUC optimization model incorporating the designed ambiguity set (M5) demonstrates greater economic efficiency compared to models using the existing nested ambiguity set and the disjoint layered ambiguity set.

Table 2.4 Comparison of optimal solutions in M3-M5, reprinted from Ref. [25]. Copyright 2024, with permission IET Renewable Power Generation

	Total cost ($)	Start up and shut-down cost ($)	Generation cost ($)	Risk cost ($)
M3	5.60×10^4	1.60×10^3	5.00×10^4	4.31×10^3
M4	5.39×10^4	1.83×10^3	4.48×10^4	7.33×10^3
M5	4.85×10^4	2.26×10^3	4.38×10^4	2.24×10^3

To further validate the effectiveness of the designed enhanced disjoint layered ambiguity set, we compare the worst-case RGFE distributions across the ambiguity sets used in M3, M4, and M5. Specifically, we focus on the worst-case distributions for time slot 14. These distributions are illustrated in Figs. 2.4, 2.5 and 2.6, where sub-uncertainty sets are depicted by solid color lines and worst-case distributions are shown as gray scatter points.

From Figs. 2.4, 2.5 and 2.6, two key observations can be made. First, there are notable differences in the worst-case RGFE distributions among M3, M4, and M5. Second, the worst-case RGFE distribution in M5 is particularly effective at capturing RGFE uncertainty. Specifically, the worst-case RGFE distribution in M5 avoids extreme fluctuations and unrealistic deviations, such as wind and PV power reaching their maximum deviations. In contrast, both M3 and M4 include these extreme outcomes in their worst-case RGFE distributions, assigning a probability of 0.005 to such scenarios. However, it is important to note that these extreme outcomes have never occurred historically, with an occurrence frequency of zero. This suggests that M3 and M4 construct a conservative RGFE distribution by assigning nonzero probabilities to extreme outcomes that have never happened in practice.

Conversely, M5 aligns with historical data by assigning zero probability to these extreme outcomes. This is because the enhanced sub-uncertainty sets in M5 are formulated using historical data information. The boundaries of these enhanced sets

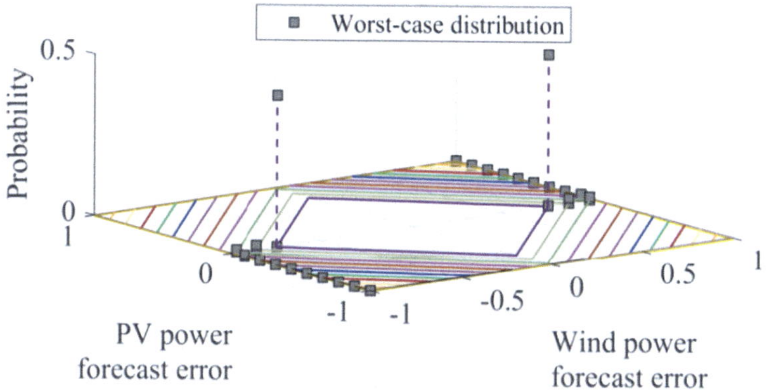

Fig. 2.4 Worst-case distributions of M3, reprinted from Ref. [25]. Copyright 2024, with permission IET Renewable Power Generation

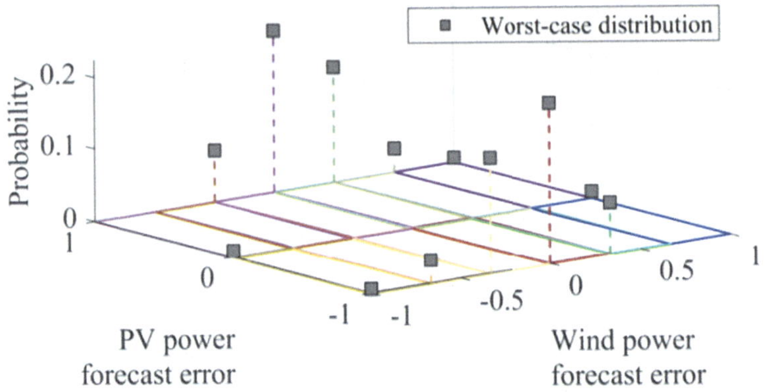

Fig. 2.5 Worst-case distributions of M4, reprinted from Ref. [25]. Copyright 2024, with permission IET Renewable Power Generation

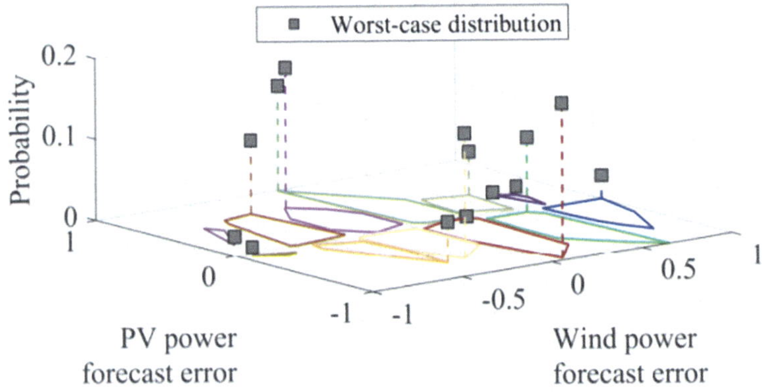

Fig. 2.6 Worst-case distributions of M5, reprinted from Ref. [25]. Copyright 2024, with permission IET Renewable Power Generation

are based on the extreme outcomes observed historically, specifically the convex hull of historical data, which ensures that the formulated RGFE distributions do not assign probability to cases outside these sets. As a result, the RGFE distribution family in M5 effectively excludes drastic fluctuations and unrealistic deviations. This demonstrates that M5 constructs a more precise and realistic RGFE distribution family compared to M3 and M4.

In summary, the designed enhanced disjoint layered ambiguity set in M5, by effectively utilizing historical data, enables the formulation of a more accurate and realistic RGFE distribution family. This improved representation of uncertainty results in more economically efficient and reliable UC decisions.

Table 2.5 Comparison of M5-M6, reprinted from Ref. [25]. Copyright 2024, with permission IET Renewable Power Generation

	Total cost ($)	Computation time (s)	Iteration number
M5	4.85×10^4	exceeds 3.60×10^3	1
M6	4.85×10^4	1.38×10^2	1

2.6.5 Validity of Proposed Worst-Case Support Set Identification Method

In this subsection, we perform simulations using M5 and M6 to evaluate the effectiveness of the proposed worst-case support set identification method. Both M5 and M6 use the enhanced disjoint layered ambiguity set, apply the novel reformulation approach, assume quadratic generation costs, and reformulate the TDRUC model as a large-scale MIQP. In M5, the complex optimization model is solved using the Gurobi solver, whereas in M6, the proposed worst-case support set identification method is employed. The results are summarized below.

Based on the results presented in Table 2.5, both M5 and M6 yield identical total costs of $\$4.85 \times 10^4$, confirming that both methods achieve the optimal solution. Furthermore, both methods converge within the first iteration, indicating their effectiveness in finding the optimal solution. However, M5 requires a computation time of 3.60×10^3 s, whereas M6 completes in approximately 1.38×10^2 s. Notably, M6, utilizing the proposed worst-case support set identification method, demonstrates a significantly lower computation time and fewer iterations, underscoring its superior computational efficiency compared to M5.

To further validate the effectiveness of the proposed solution method (M6), we compare the worst-case RGFE distribution support sets obtained from M5 and M6. For clarity, we focus on the support sets in time slot 14, as illustrated in Figs. 2.7 and 2.8. The support sets are depicted as gray scatter points.

The worst-case distribution and the corresponding worst-case distribution support set for M5 are illustrated in Figs. 2.6 and 2.7, respectively. These figures reveal that only a small subset of vertexes within the sub-uncertainty sets are critical for determining the worst-case RGFE distribution. This finding is consistent with Theorem 2, which asserts that these pivotal vertexes can be identified using a linear program. Consequently, the MIQP model's original complexity in M5 can be reduced by removing constraints associated with inactive vertexes.

Figure 2.8 highlights the effectiveness of the worst-case support set identification method used in M6. By focusing on the essential vertexes identified from the worst-case distribution support set, M6 efficiently relaxes the original MIQP model, thereby enhancing computational efficiency. As a result, M6 achieves a significant reduction in computation time compared to M5. In summary, the proposed worst-case support set identification method effectively facilitates the optimal solution of the TDRUC optimization model in a computationally efficient manner.

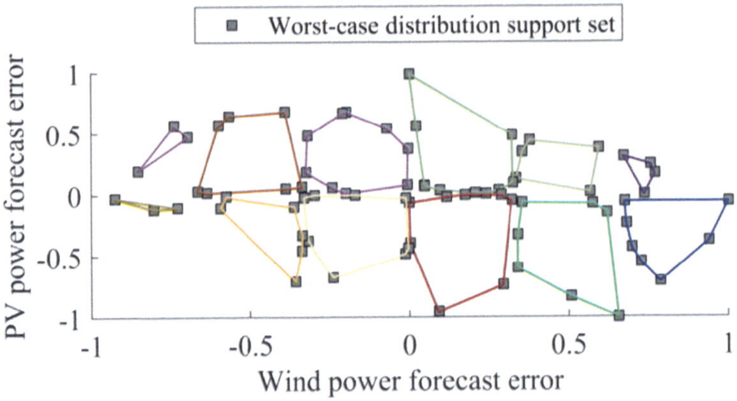

Fig. 2.7 Worst-case distribution support set of M5, reprinted from Ref. [25]. Copyright 2024, with permission IET Renewable Power Generation

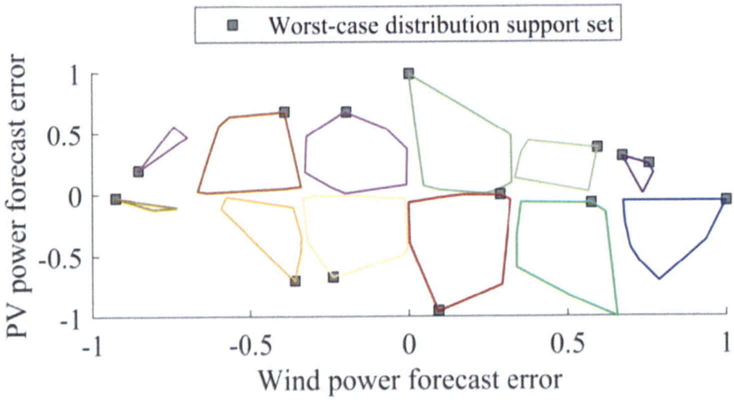

Fig. 2.8 Worst-case distribution support set of M6, reprinted from Ref. [25]. Copyright 2024, with permission IET Renewable Power Generation

2.7 Conclusion

In this chapter, an enhanced disjoint layered ambiguity set is introduced to improve UC decisions within the DRO framework. Unlike existing nested or disjoint layered ambiguity sets, the proposed set leverages historical data to define more accurate sub-uncertainty sets, effectively excluding unrealistic deviation events of renewable generation (RGs) and resulting in a more precise distribution family. This approach is integrated with a novel reformulation technique for the TDRUC optimization model and a worst-case support set identification method, ensuring that UC decisions are both effective and computationally efficient.

Simulation results highlight the advantages of using the enhanced disjoint layered ambiguity set, novel reformulation approach, and worst-case support set identification method under high renewable power penetration conditions. The proposed TDRUC model, which employs the enhanced ambiguity set, demonstrates superior economic performance and reliability compared to models based on existing nested or disjoint layered ambiguity sets. Additionally, the simulation results reveal that the novel reformulation approach significantly improves computational efficiency over the traditional Benders decomposition method, reducing solution time. The worst-case support set identification method further simplifies model complexity, enhancing the applicability and practicality of the TDRUC model. In conclusion, the integration of the enhanced disjoint layered ambiguity set, novel reformulation approach, and worst-case support set identification method enables the achievement of economically and reliably optimized UC decisions in a computationally efficient manner.

References

1. Child M, Breyer C (2016) Vision and initial feasibility analysis of a recarbonised finnish energy system for 2050. Renew Sustain Energy Rev 66:517–536
2. Tigas K, Giannakidis G, Mantzaris J, Lalas D, Sakellaridis N, Nakos C, Vougiouklakis Y, Theofilidi M, Pyrgioti E, Alexan-dridis A (2015) Wide scale penetration of renewable electricity in the Greek energy system in view of the European decarbonization targets for 2050 Renew Sustain Energy Rev 42:158–169
3. Kroposki B, Johnson B, Zhang Y, Gevorgian V, Denholm P, Hodge B-M, Hannegan B (2017) Achieving a 100% renewable grid: operating electric power systems with extremely high levels of variable renewable energy. IEEE Power Energy Mag 15(2):61–73
4. Qiu H, Gu W, Liu P, Sun Q, Wu Z, Lu X (2022) Application of two- stage robust optimization theory in power system scheduling under uncertainties: a review and perspective. Energy 251:123942
5. Verastegui F, Lorca A, Olivares DE, Negrete-Pincetic M, Gaz- muri P (2019) An adaptive robust optimization model for power systems planning with operational uncertainty. IEEE Trans Power Syst 34(6):4606–4616
6. Amjady N, Dehghan S, Attarha A, Conejo AJ (2016) Adaptive robust network-constrained AC unit commitment. IEEE Trans Power Syst 32(1):672–683
7. Li Z, Floudas CA (2012) Robust counterpart optimization: uncertainty sets, formulations and probabilistic guarantees. In: Proceedings of International Conference on Foundation of Computer-aided Process Operations (FOCAPO). Savannah, Georgia
8. Zhang Y, Han X, Xu B, Wang Y, Wang M, Ye P (2018) Risk-based reserve coordinative unit commitment for a large-scale wind- storage system. Electr Power Compon Syst 46(19–20):2004–2020
9. Huang H, Li Z, Gooi HB, Qiu H, Zhang X, Lv C, Liang R, Gong D (2023) Distributionally robust energy-transportation coordination in coal mine integrated energy systems. Appl Energy 333:120577
10. Zhong J, Li Y, Wu Y, Cao Y, Li Z, Peng Y, Qiao X, Xu Y, Yu Q, Yang X et al (2023) Optimal operation of energy hub: an integrated model combined distributionally robust optimization method with stackelberg game. IEEE Trans Sustain Energy
11. Xiong P, Jirutitijaroen P, Singh C (2016) A distributionally robust optimization model for unit commitment considering uncertain wind power generation. IEEE Trans Power Syst 32(1):39–49

12. Zheng X, Chen H, Xu Y, Li Z, Lin Z, Liang Z (2020) A mixedinteger SDP solution to distributionally robust unit commitment with second order moment constraints. CSEE J Power Energy Syst 6(2):374–383
13. Zheng X, Qu K, Lv J, Li Z, Zeng B (2020) Addressing the conditional and correlated wind power forecast errors in unit commitment by distributionally robust optimization. IEEE Trans Sustain Energy 12(2):944–954
14. Du Y, Jiang L, Duan C, Li Y, Smith J (2017) Energy consumption scheduling of HVAC considering weather forecast error through the distributionally robust approach. IEEE Trans Ind Inform 14(3):846–857
15. Wang Y, Du Y, Duan C, Xu H, Jiang L (2019) Data-driven distributionally robust energy consumption scheduling of HVAC based on disjoint layered ambiguity set. In: 2019 IEEE Power and Energy Society General Meeting (PESGM). IEEE, pp 1–5
16. Wang Y, Zeng Q, Tian Z, Du Y, Zeng P, Jiang L (2022) Distributionally robust energy consumption scheduling of hvac considering the uncertainty of outdoor temperature and human activities. CSEE J Power Energy Syst
17. Xiong P, Singh C (2018) A distributional interpretation of uncertainty set in unit commitment under uncertain wind power. IEEE Trans Sustain Energy 10(1):149–157
18. Zhang Y, Liu W, Huang Z, Zheng F, Le J, Zhu S (2021) Distributionally robust coordination optimization scheduling for electricity-gastransportation coupled system considering multiple uncertainties. Renew Energy 163:2037–2052
19. Gong H, Zhang ZH (2022) Benders decomposition for the distributionally robust optimization of pricing and reverse logistics network design in remanufacturing systems. Eur J Oper Res 297(2):496–510
20. Chen Y, Guo Q, Sun H, Li Z, Wu W, Li Z (2018) A distributionally robust optimization model for unit commitment based on kullback–leibler divergence. IEEE Trans Power Syst 33(5):5147–5160
21. Wang S, Zhao C, Fan L, Bo R (2022) Distributionally robust unit commitment with flexible generation resources considering renewable energy uncertainty. IEEE Trans Power Syst 37(6):4179–4190
22. Alismail F, Xiong P, Singh C (2017) Optimal wind farm allocation in multi-area power systems using distributionally robust optimization approach. IEEE Trans Power Syst 33(1):536–544
23. Lowery C, O'Malley M (2012) Impact of wind forecast error statistics upon unit commitment. IEEE Trans Sustain Energy 3(4):760–768
24. Che L, Liu X, Zhu X, Cui M, Li Z (2018) Intra-interval security assessment in power systems with high wind penetration. IEEE Trans Sustain Energy 10(4):1890–1903
25. Lian Y, Li Y, Zhao Y et al (2024) A computational efficient approach for distributionally robust unit commitment with enhanced disjointed layered ambiguity set. IET Renew Power Gener
26. Velloso A, Street A, Pozo D, Arroyo JM, Cobos NG (2019) Two-stage robust unit commitment for co-optimized electricity markets: an adaptive data-driven approach for scenario-based uncertainty sets. IEEE Trans Sustain Energy 11(2):958–969
27. Ding Z, Xie L, Lu Y, Wang P, Xia S (2018) Emission-aware stochastic resource planning scheme for data center microgrid considering batch workload scheduling and risk management. IEEE Trans Ind Appl 54(6):5599–5608
28. Zhang J, Hodge B-M, Florita A (2015) Joint probability distribution and correlation analysis of wind and solar power forecast errors in the western interconnection. J Energy Eng 141(1):B4014008
29. Li P, Yu D, Yang M, Wang J (2018) Flexible look-ahead dispatch realized by robust optimization considering CVaR of wind power. IEEE Trans Power Syst 33(5):5330–5340
30. Lofberg J (2004) YALMIP: a toolbox for modeling and optimization in MATLAB. In: 2004 IEEE international conference on robotics and automation (IEEE Cat. No. 04CH37508). IEEE, pp 284–289
31. Gurobi Optimization L (2018) Gurobi optimizer reference manual

Chapter 3
Downside Risk to Low-Carbon Multi-energy System Optimization

3.1 Introduction

In recent years, the escalating challenges of environmental pollution and energy shortage have gained significant global attention. Addressing carbon dioxide emissions and transitioning to a low-carbon way are critical for ensuring the long-term sustainable development [1]. Historically, the world's economic development has been deeply dependent on coal, with the power industry being a major contributor to fossil consumption and associated carbon emissions [2]. In response to this, Integrated Energy Systems (IES) have emerged as a promising technology due to their potential for higher efficiency and carbon emission reduction [3].

A core principle of IES technology is multi-energy complementarity, which enhances overall system efficiency. However, this approach alone may not sufficiently resolve the carbon emissions issue [4]. Thus, there is an urgent need to explore effective low-carbon strategies specifically designed for IES in the context of sustainable development [5]. This need underscores the shift toward greener and more sustainable energy systems, emphasizing the importance of innovative technologies and strategic planning in reducing the environmental impacts of energy production and consumption.

Among the various low-carbon strategies, carbon capture and the utilization of renewable energy are crucial for advancing sustainable energy solutions. Notably, Zhang et al. [6] proposed a collaborative framework for IES that integrates carbon considerations, aiming to clarify the complex relationship between emission reduction and the deployment of renewable energy resources. This innovative framework incorporates carbon capture technique, extending its application to the broader context of IES, which includes electricity, thermal, and gas components [7].

Furthermore, the emergence of market-driven trading mechanisms has become a pivotal incentive to accommodate renewable energy sources and reduce carbon emissions. Extensive research has explored various aspects of carbon trading (CT). For instance, Wang et al. [8] developed a two-stage stochastic planning model for IES

J. Chen and Y. Li, *Renewable Power System Optimization*, Smart Energy Systems,
https://doi.org/10.1007/978-981-97-8132-4_3

that incorporates carbon emission pricing, aiming to identify optimal planning strategies under different CT frameworks. Similarly, Li et al. [9] introduced an innovative approach that integrates CT mechanisms to IES planning to achieve low-carbon operations.

However, it is important to note that existing research often lacks detailed quantification of carbon trading and capture in practical scenarios, particularly concerning deviations in power outputs from renewable power integrated fossil energy generations [10]. In low-carbon IES, the uncertainty surrounding renewable power outputs is critical, as it influences both carbon emissions and operational risks for IES frameworks [11]. To address this challenge, two commonly employed methods are robust optimization and stochastic optimization [12]. Robust optimization, which focuses on worst-case scenarios to ensure system reliability across diverse conditions, is often criticized for its conservatism [13]. While it is effective in maintaining system performance under extreme conditions, its overly cautious nature can impede optimal performance in more typical scenarios. This highlights the need for methodologies that balance robustness with operational efficiency in low-carbon IES frameworks.

Stochastic optimization has become a prominent technique for addressing uncertainty across various domains, utilizing random sampling methods to generate solutions. Within this field, two widely adopted approaches are the conditional value-at-risk (CVaR) and mean–variance (MV) model. The MV model, introduced by Markowitz, represents returns through expected values and risks through variance [14]. It has gained significant traction in power system operations, particularly in integrating uncertain renewable energy sources [15]. This model is often employed in economic dispatch problems involving wind power, aiming to optimize both profitability and risk in uncertain environments [16]. In [17], a mean–variance-skewness-based multi-objective optimization model was proposed, designed to maximize expected returns and skewness while minimizing risk. Nonetheless, the multi-objective nature of the MV model can sometimes present challenges in terms of accuracy and efficiency [18].

Recent research has integrated CVaR theory with stochastic optimization, offering a robust framework for risk assessment and scheduling in virtual power plants connected to renewable energy sources [19]. This approach has also been extended to the planning of IES, providing a comprehensive risk assessment model that accounts for uncertainties in energy prices, with CVaR serving as the risk measure [20]. CVaR provides a nuanced perspective on risk assessment [21]. However, the subjectivity involved in defining expectations and risk levels continues to be a point of debate regarding the accuracy of the CVaR model for uncertainty measurement [22].

Numerous models within probability theory have been developed to describe uncertainties as random variables. But practical engineering problems often involve non-probabilistic uncertainties characterized by vagueness and ambiguity, which present challenges in renewable power integrated system optimization [23, 24]. To address these issues, fuzzy set theory has emerged as a valuable framework for managing uncertainties arising from renewable energy sources [25]. For instance, Miranda et al. [26] utilized fuzzy measures to study uncertain wind power and sought solutions that balance its penetration with variations in optimization objective. Additionally, a fuzzy scheduling was proposed to address multi-uncertainty in IES [27]. Furthermore, a multi-objective scheduling approach, employing a fuzzy surrogate worth trade-off method, was introduced to tackle economic, and emission challenges [28].

Despite its advantages, traditional fuzzy theory encounters limitations, particularly concerning the self-duality of membership functions, which may lead to inconclusive results [29]. This ambiguity can complicate decision-making processes, potentially resulting in suboptimal outcomes [30]. Therefore, there is an increasing need to develop models to capture the stochastic characteristics of renewable energy in addressing the fuzzy nature of uncertainties, considering incomplete human knowledge influenced by weather, environmental factors, and operational conditions [31]. The convergence of stochastic and fuzzy optimization methods may offer a promising approach to address these complex challenges in renewable power system optimization.

In addressing the aforementioned challenges, the entropy model has proven to be a valuable tool for managing uncertainty in fuzzy-random environments [32]. The seminal work linking entropy with risk measurement in portfolio optimization was introduced by [33], which defined entropy as a means to capture uncertainty arising from the ambiguity of potential fuzzy variables, by developing a credibility measure. In fact, entropy effectively encapsulates the uncertainty surrounding both lower and higher down side returns, adopting a conservative approach to address the risk in the context of uncertain renewable energy, though this may come at the expense of potentially higher returns. As a result, entropy does not differentiate between returns and possible losses in the operation of IES. For IES operators, significant risk materializes when returns fall below the expected value, indicating that the operational costs of the IES exceed anticipated levels.

Despite this, few studies have specifically addressed situations where IES operational costs surpass expected values. To bridge this gap, this chapter introduces the concept of semi-entropy as a measure of downside risk, representing scenarios where operational costs exceed expectations. Based on credibility theory, a novel mean-semi-entropy is then developed, aiming to balance economic efficiency and risks raised by uncertain wind power within an IES.

In power systems incorporating wind farms, it is often assumed that power outputs are independent across different turbines [34, 35]. However, this assumption overlooks the coupling and dependence among different wind turbines, which can lead to increased uncertain risks because of inaccurate power flow calculations [36]. Previous studies, such as [37] and [38], have utilized Cholesky decomposition and linear regression to model correlations between wind turbines. Linear regression provides a robust framework for capturing linear relationships between uncertain variables, while Cholesky decomposition with covariance matrix is a widely used linear algebra method. Both methods, however, are limited to linear correlations and may not fully capture the complexity of wind power dependencies.

To address these limitations, a novel approach based on the Nataf transformation—a type of copula technique—has been proposed [39]. This approach has rapidly gained acceptance for short-term dispatch in assuming forecasting errors follow a Gaussian distribution [40, 41]. Obtaining the joint distribution of random variables is the most direct method to fully describe their correlations. In this context, the multivariate Weibull has been proposed to model the joint distribution across different turbines, although this assumption may not always align perfectly with real-world data [42]. The Gaussian mixture model has also been employed as an effective tool for fitting the joint distribution [43]. However, multivariate parameter estimation becomes increasingly challenging with high-dimensional random inputs. To address this, the multivariate Gaussian Copula has been introduced as an advanced estimation technique to elucidate the wind farm output correlations among multiple wind turbines.

The subsequent sections of this chapter are organized as follows: Section 3.2 provides a detailed system description and mathematical modeling. The concept of semi-entropy for uncertain renewable energy is defined in Sect. 3.3. Following this, Sect. 3.4 presents low-carbon IES optimization model considering downside risk. Case studies are then showcased in the subsequent section. Finally, Sect. 3.6 concludes the chapter, summarizing the key findings and implications of the study.

3.2 Low-Carbon Integrated Energy System

Figure 3.1 illustrates an IES schematic. The primary purpose of this work is to achieve cooperative optimization within an integrated electricity-gas-heat energy system considering low-carbon factor. This system integrates several energy parts, consisting of coal-fired unit, natural gas-fired unit, CHP units, wind turbines, as well as electric boilers (EBs).

3.2.1 Carbon Capture System

At time t, assuming the total carbon emission volume is Q_t^{CEV} and the carbon capture system (CCS) model can be expressed as follows:

$$Q_t^{\mathrm{CEV}} = Q_t^{\mathrm{CEV},C} + Q_t^{\mathrm{CEV},G} \tag{3.1}$$

$$Q_t^{\mathrm{CEV},C} = \alpha^{\mathrm{PG},C}(P_t^{\mathrm{PG},C} + \Delta P_t^{\mathrm{PG},C})\Delta t \tag{3.2}$$

$$Q_t^{\mathrm{CEV},G} = \alpha^{\mathrm{PG},G}(P_t^{\mathrm{PG},G} + \Delta P_t^{\mathrm{PG},G})\Delta t \tag{3.3}$$

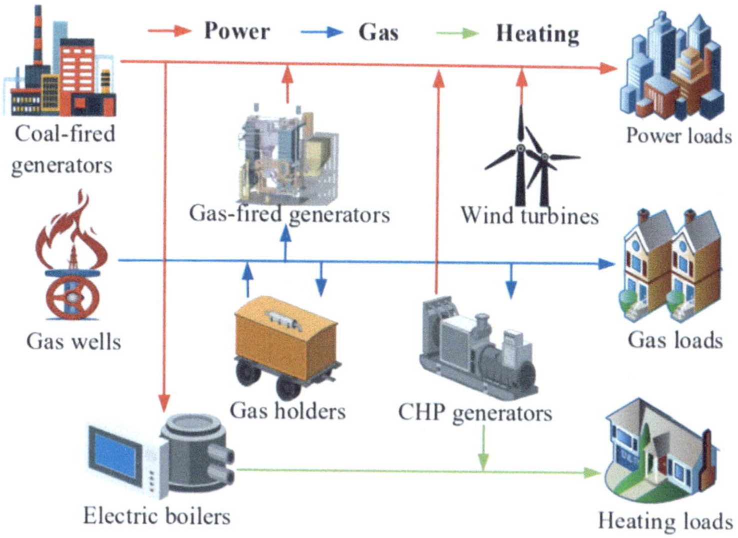

Fig. 3.1 Overall structure of IES, reprinted from Ref. [57]. Copyright 2024, with permission Energy

where $Q_t^{\mathrm{CEV},G}$ and $Q_t^{\mathrm{CEV},C}$ denote the carbon emission of gas-fired and coal-fired, respectively, $P_t^{\mathrm{PG},G}$ and $P_t^{\mathrm{PG},C}$ denote gas-fired and coal-fired outputs at time t, respectively, $\alpha^{\mathrm{PG},G}$ and $\alpha^{\mathrm{PG},C}$ represent emission intensity of gas-fired and coal-fired generators, respectively, $\Delta P_t^{\mathrm{PG},C}$ and $\Delta P_t^{\mathrm{PG},G}$ indicate the deviations of coal-fired and gas-fired outputs due to uncertain wind power, respectively, $\Delta P_t^{\mathrm{PG},C} \in \left[-\Delta P_{\max}^{\mathrm{PG},C}, \Delta P_{\max}^{\mathrm{PG},C}\right]$, $\Delta P_t^{\mathrm{PG},G} \in \left[-\Delta P_{\max}^{\mathrm{PG},G}, \Delta P_{\max}^{\mathrm{PG},G}\right]$, $\Delta P_{\max}^{\mathrm{PG},G}$ and $\Delta P_{\max}^{\mathrm{PG},C}$ are the maximum deviations of coal-fired and gas-fired outputs, respectively, and the time interval Δt is set to 1 h.

As for the cost of CCS, it involves carbon storage and transmission cost:

$$f_{\mathrm{CCS}} = \sum\nolimits_{t=1}^{T} \eta Q_t^{\mathrm{CEV}} \tag{3.4}$$

where η denotes the unit CCS cost, T represents the total number of time scale.

Additionally, the carbon captured from the CCS must satisfy:

$$0 \le Q_t^{\mathrm{CEV}} \le Q_{t,\max}^{\mathrm{CEV}} \tag{3.5}$$

$$Q_t^{\mathrm{CEV}} - \underline{QR} \le Q_{t+1}^{\mathrm{CEV}} \le Q_t^{\mathrm{CEV}} + \overline{QR} \tag{3.6}$$

where $Q_{t,\max}^{\mathrm{CEV}}$ equals to $\alpha_{\max} Q_t^{\mathrm{CEV}}$, $\alpha_{\max}$ denotes the maximum emission intensity, $\overline{QR}$ and $\underline{QR}$ represent ramp limits of CCS.

3.2.2 Carbon Trading

In the context of CT, the free allocation method is commonly employed to distribute carbon credits within industry. We assume that an operator oversees IES operation. Once the emissions from coal-fired and gas-fired generators exceed the allocated carbon quota, the operator must purchase additional emission quotas within the IES, incurring CT costs. Conversely, if the carbon emissions from these generators are lower than the allocated value because of substantial wind power utilization, the operator will generate revenue by using CT. This mechanism incentivizes the increased utilization of wind power.

The allocated carbon quotas for coal-fired and gas-fired are:

$$Q_t^{\mathrm{CQ}} = \delta_C \sum\nolimits_{i=1}^{N_C} \left(P_t^{\mathrm{PG},C} + \Delta P_t^{\mathrm{PG},C}\right) \tag{3.7}$$

$$Q_t^{\mathrm{GQ}} = \delta_G \sum\nolimits_{i=1}^{N_G} \left(P_t^{\mathrm{PG},G} + \Delta P_t^{\mathrm{PG},G}\right) \tag{3.8}$$

where Q_t^{GQ} and Q_t^{CQ} denote the allocated carbon quota of gas-fired and coal-fired, respectively, δ_C and δ_G represent electricity carbon conversion coefficients, N_G and N_C indicate the natural gas-fired and coal-fired generators, respectively. P_t^C and P_t^G are coal-fired and natural gas-fired outputs, respectively.

The CT cost f_{CT} can be determined with the CT price χ_{CT}:

$$f_{\mathrm{CT}} = \sum\nolimits_{t=1}^{T} \chi_{\mathrm{CT}}(Q_t^{\mathrm{CEV},C} - Q_t^{\mathrm{CQ}} + Q_t^{\mathrm{CEV},G} - Q_t^{\mathrm{GQ}}) \tag{3.9}$$

3.2.3 IES Objective

To achieve low-carbon and economic operation, the optimization objective of the proposed IES is formulated as follows:

$$OC = f_{\mathrm{CCS}} + f_{\mathrm{CT}} + f_{\mathrm{GC}} + f_{\mathrm{GW}} + f_{\mathrm{CHP}} \tag{3.10}$$

where

$$f_{\mathrm{GC}} = \sum\nolimits_{t=1}^{T} \left[\sum\nolimits_{i=1}^{N_{\mathrm{GS}}} \left(a_i^{\mathrm{GS}} + b_i^{\mathrm{GS}} P_{i,t}^{\mathrm{GS}} + c_i^{\mathrm{GS}} P_{i,t}^{\mathrm{GS2}}\right)\right] \tag{3.11}$$

$$f_{\mathrm{GW}} = \sum_{t=1}^{T} \left[\chi_{\mathrm{GW}} \sum\nolimits_{i=1}^{N_{\mathrm{GW}}} G_{i,t}^{\mathrm{GW}}\right] \tag{3.12}$$

$$f_{\text{CHP}} = \sum_{t=1}^{T} \left[\sum_{i=1}^{N_{\text{CHP}}} \chi_{\text{CHP}} \left(\gamma_P P_{i,t}^{\text{CHP}} + \gamma_{\text{H}} H_{i,t}^{\text{CHP}} \right) \right] \tag{3.13}$$

where $P_{i,t}^{\text{GS}}$ denotes the output of the ith generator at time t, a_i^{GS}, b_i^{GS} and c_i^{GS} indicate cost coefficients. In Eq. (3.12), f_{GW} is the gas production cost, where N_{GW} denotes the number of gas wells, $G_{i,t}^{\text{GW}}$ denotes ith gas well production at time t, and χ_{GW} represent gas price. Equation (3.13) gives CHP operation cost, where N_{CHP} denotes the number of CHP, $P_{i,t}^{\text{CHP}}$ and $H_{i,t}^{\text{CHP}}$ indicate the electrical power and heat power generated by CHP i during time t, respectively, χ_{CHP} represents CHP cost coefficient, γ_P and γ_H denote the conversion coefficients describing the fuel consumed by CHP's electrical power and the thermal power, respectively.

3.2.4 Electricity Network Constraints

The power flow constraint of IES can be modeled as:

$$PF_{l,t} = \left(\theta_{v,t} - \theta_{w,t} \right) / Im_l \tag{3.14}$$

where $\theta_{v,t}$, $\theta_{w,t}$ represent the voltage angle of nodes v and w at time t, respectively, and Im_l denotes line l impedance.

The operational limits for generators, which encompass output limit, ramp-up limit, and ramp-down limit, are defined as follows:

$$P_{i,\min}^{\text{GS}} \le P_{i,t}^{\text{GS}} \le P_{i,\max}^{\text{GS}} \tag{3.15}$$

$$P_{i,t}^{\text{GS}} - P_{i,t-1}^{\text{GS}} \le R_{U,i} \tag{3.16}$$

$$P_{i,t-1}^{\text{GS}} - P_{i,t}^{\text{GS}} \le R_{D,i} \tag{3.17}$$

where $P_{i,\min}^{\text{GS}}$ and $P_{i,\max}^{\text{GS}}$ denote the minimum and maximum limits of $P_i^{\text{GS}} \in \{P_i^C, P_i^W, P_i^G, P_i^{\text{CHP}}, P_i^{\text{EB}}\}$, $R_{U,i}$ and $R_{D,i}$ represent the ramp-up and ramp-down constraints of the ith generator, respectively.

The power balance constraint in IES can be expressed as:

$$P_t^C + P_t^G + P_t^W + P_t^{\text{CHP}} + PF_t - P_t^D - P_t^{\text{EB}} = 0 \tag{3.18}$$

where P_t^D, P_t^W, and P_t^{CHP} represent load demand, wind turbine output, and CHP output at time t, respectively.

3.2.5 Natural Gas Network Constraints

In IES, the natural gas network includes gas wells, pipelines, compressors, storage facilities, and gas loads, which is illustrated in Fig. 3.2.

The gas well production $G_{i,t}^{\mathrm{GW}}$ and nodal pressure $p_{k,t}$ are subject to the following constraints:

$$G_{i,\min}^{\mathrm{GW}} \leq G_{i,t}^{\mathrm{GW}} \leq G_{i,\max}^{\mathrm{GW}} \tag{3.19}$$

$$p_k^{\min} \leq p_{k,t} \leq p_k^{\max} \tag{3.20}$$

where $G_{i,\min}^{\mathrm{GW}}$ and $G_{i,\max}^{\mathrm{GW}}$ are the minimum and maximum production of gas well i, respectively, $p_k^{\min}$ and $p_k^{\max}$ represent the lower and upper pressures at node k, respectively.

The gas flow constraint can then be formulated as follows:

$$G_{k,n,t}^{\mathrm{GF}} = c_{k,n}\sqrt{\left|p_{k,t}^2 - p_{n,t}^2\right|} \tag{3.21}$$

where $G_{k,n,t}^{\mathrm{GF}}$ denotes the gas flow from nodes k to n at time t, and $c_{k,n}$ represents pipeline parameter.

Additionally, the outlet pressure $p_{j,t}$ associated with the compressor and inlet pressure $p_{n,t}$ holds $p_{j,t} = \rho_t \times p_{n,t}$, where ρ_t denotes the compressor compression ratio.

The gas holder, which functions as a storage equipment for natural gas, is formulated as:

$$G_t^S = G_{t-1}^S + G_t^{\mathrm{in}} - G_t^{\mathrm{out}} \tag{3.22}$$

$$0 \leq G_t^S \leq G_{\max}^S \tag{3.23}$$

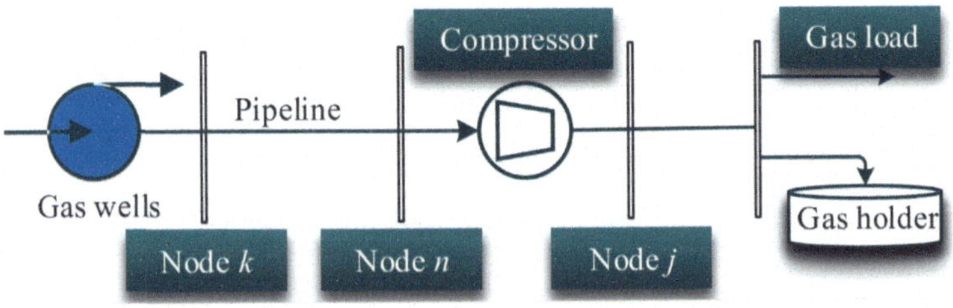

Fig. 3.2 Structure of natural gas network, reprinted from Ref. [57]. Copyright 2024, with permission Energy

where G_t^S and G_{t-1}^S denote stored gas of holder during time t and $t-1$, respectively, G_t^{out} and G_t^{in} denote the natural gas that outflow from and inflow to gas holder, respectively.

The gas balance constraint is presented as:

$$G_t^{\text{GW}} + G_t^{\text{GF}} - G_t^{\text{GAS}} - G_t^G - G_t^{\text{CHP}} = 0 \tag{3.24}$$

where G_t^{GAS}, G_t^G, and G_t^{CHP} are the gas load, the consumption of gas-fired, and CHP generators during time t, respectively.

Equation (3.21) presents a non-linear and non-convex constraint. To address this, the linearization technique is developed to piecewise linearize Eq. (3.21), facilitating the transformation of the proposed model into a mixed-integer linear programming (MILP). Detailed information on the linearization technique can be found in [44] and is not elaborated upon here.

3.2.6 Heating Network Constraints

The electricity consumption P_t^{EB} and heat production H_t^{EB} of electric boiler at time t is formulated by Eq. (3.25). Similarly, the electricity consumption P_t^{CHP} and heat production H_t^{CHP} of CHP at time t can be modeled by Eq. (3.26).

$$H_t^{\text{EB}} = \in^{\text{EB}} P_t^{\text{EB}} \tag{3.25}$$

$$H_t^{\text{CHP}} = \in^{\text{CHP}} P_t^{\text{CHP}} \tag{3.26}$$

where $\in^{\text{EB}}$ and $\in^{\text{CHP}}$ are the conversion efficiencies related to EB and CHP, respectively.

Here, the supply and return water structure of heating network is illustrated in Fig. 3.3. The nodal temperatures under mass flow rates can be expressed as follows:

$$\begin{aligned} T_{\text{ms,in}} ms_{\text{in}} &= T_{\text{ms,out}} ms_{\text{out}} \\ T_{\text{mr,in}} mr_{\text{in}} &= T_{\text{mr,out}} mr_{\text{out}} \end{aligned} \tag{3.27}$$

where $T_{\text{ms,in}}$, $T_{\text{ms,out}}$, $T_{\text{mr,in}}$, and $T_{\text{mr,out}}$ represent the nodal temperature of supply and return pipes before and after the water mixes, respectively, satisfying $T_{\text{ms,out}} = T_{\text{ms}}$, $T_{\text{mr,out}} = T_{\text{mr}}$, T_{ms} and T_{mr} are nodal mixing temperatures related to the supply and return pipes, respectively. ms_{in}, ms_{out}, mr_{in}, and mr_{out} indicate the mass flow rates of supply and return pipes before and after the water mixes, respectively.

The heat load H_t^D and temperature satisfy the following equation:

$$H_t^D = C_{\text{p}} m_t (T_{\text{ms}} - T_{\text{mr}}) \tag{3.28}$$

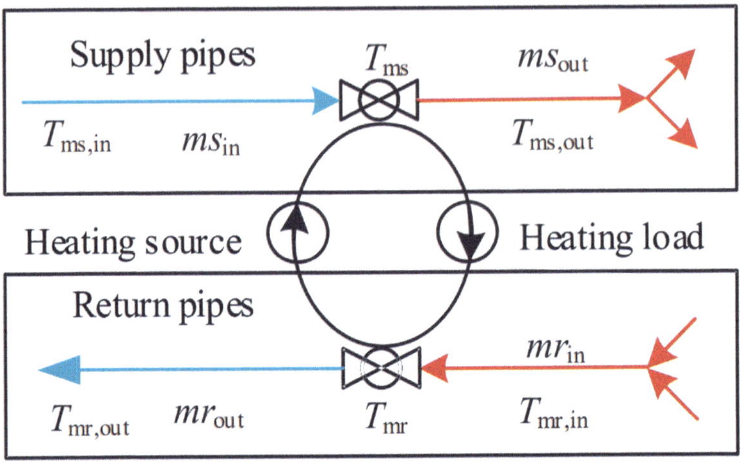

Fig. 3.3 Heating network diagram, reprinted from Ref. [57]. Copyright 2024, with permission Energy

where C_p and m_t denote the heat capacity and the amount of injected water at time t, respectively.

The heating network balance constraint is:

$$H_t^D - H_t^{\text{CHP}} - H_t^{\text{EB}} = 0 \tag{3.29}$$

3.3 Fuzzy Semi-entropy-Based Downside Risk Model

To quantify the likelihood of fuzzy variable, Zadeh introduced concepts of necessity and possibility measures in [45, 46]. While these measures satisfy normality, monotonicity and non-negativity properties, the dual was missed; that is, the sum of the possibility and necessity measures for a fuzzy variable and its complement does not necessarily equal to one [32]. In view of this, entropy based on credibility measures was proposed to model the uncertainty arising from fuzzy variables [32]. Research indicates that entropy, when used to simulate risk measures, provides a more robust assessment of uncertainty compared to variance, thereby improving wealth allocation strategies [33].

Entropy captures the uncertainty associated with both high and low extreme returns, adopting a conservative stance in addressing risks related to uncertain wind power, which may result in sacrificing potential higher returns. The critical uncertainty affecting IES operation is the risk of returns falling below the expected value.

To address this, this chapter introduces a mean-semi-entropy (MSE) model with credibility theory for evaluating the downside risk in uncertain wind power integrated IES.

Initially, a triangular function is used to represent the fuzzy characteristics of uncertain wind power. Consequently, the triangular function for the wind output can be expressed as:

$$\mu(P^W) = \begin{cases} \frac{P^W - P^W_{\min}}{P^W_{\text{mid}} - P^W_{\min}}, & P^W_{\min} \le P^W \le P^W_{\text{mid}} \\ \frac{P^W_{\max} - P^W}{P^W_{\max} - P^W_{\text{mid}}}, & P^W_{\text{mid}} \le P^W \le P^W_{\max} \\ 0 & \text{, otherwise} \end{cases} \tag{3.30}$$

where $P^W_{\min}$, P^W_{mid}, and $P^W_{\max}$ are respectively the minimum, middle and maximum forecasting values of wind power.

Let $\xi \in \{x_1, x_2, \cdots, x_N\}$ be the fuzzy variable and its membership function is μ, x_i represents the output of ith wind power sampling in N samples. The entropy of this fuzzy variable is defined as:

$$H[\xi] = \sum_{i=1}^{N} S(\nu(x_i)) \tag{3.31}$$

where $\nu(x_i) = \left(\sup_{x_i \in A} \mu(x_i) + 1 - \sup_{x_i \in A^c} \mu(x_i) \right) / 2$ represents the credibility function, and $S(\nu) = -\nu \ln \nu - (1 - \nu) \ln(1 - \nu)$.

Subsequently, the semi-entropy is introduced to evaluate the downside risk, and it is written as:

$$S_h[\xi] = \sum_{i=1}^{N} S(\nu(x_i)^-) \tag{3.32}$$

$$\nu(x_i)^- = \begin{cases} \nu(x_i), & \text{if } OC \ge e \\ 0, & \text{otherwise} \end{cases} \tag{3.33}$$

where e denotes the expected value related to the fuzzy variable ξ.

Due to $S(0) = 0$, the semi-entropy is expressed as:

$$S_h[\xi] = \sum_{OC \ge e} S(\nu(x_i)) \tag{3.34}$$

$\xi = (a, b, c)$ indicates a triangular variable and its expected value is $e = (a + 2b + c)/4$. Then, we have the semi-entropy:

$$S_h[\xi] = \begin{cases} (b-a)\rho - (b-a)\zeta(\rho), & \text{if } e \le b \\ \frac{b-a}{2} + (c-b)\zeta(\tau), & \text{otherwise} \end{cases} \tag{3.35}$$

where $\rho = (2b + c - 3a)/8(b - a)$, $\tau = (3c - 2b - a)/8(c - b)$, then $\zeta(\rho) = \rho^2 \ln\rho - (1-\rho)^2 \ln\rho - (1-\rho)$ and $\zeta(\tau) = \tau^2 \ln\tau - (1-\tau)^2 \ln\tau - (1-\tau)$.

Moreover, if ξ is a symmetric triangular variable, which has $b - a = c - b$, the value of semi-entropy is $(b - a)/2$.

3.4 Multi-energy System Optimization with Fuzzy Semi-entropy

Wind turbines are typically clustered in regions with abundant wind resources, necessitating consideration of the correlation between wind outputs from multi-turbines, particularly in the context of wind farm large-scale access. In view of this, a multivariate Gaussian Copula (MGC) is proposed to model power outputs of multiple wind turbines and to facilitate investigate the correlation among different wind turbines [47]. For a vector $w = [w_1, \ldots, w_d]$ representing the power outputs from d wind turbines, their joint distribution M is derived by combining the marginal distribution functions with a unique Copula function C. Suppose the cumulative distribution function (CDF) and probability density function (PDF) are respectively $u_i = F(w_i)$ and m, that is

$$\begin{aligned} M(w_1, \ldots, w_d) &= C(F(w_1), \ldots, F(w_d)) \\ C(u_1, \ldots, u_d) &= M\left(F^{-1}(u_1), \ldots, F^{-1}(u_d)\right) \end{aligned} \tag{3.36}$$

$$m(w_1, w_2, \ldots, w_d) = \frac{\partial^d C(u_1, \ldots, u_d)}{\partial u_1 \ldots \partial u_d} \frac{\partial F(w_1)}{\partial w_1} \cdots \frac{\partial F(w_d)}{\partial w_d} \tag{3.37}$$

Let A be the covariance matrix related to power outputs from different wind turbines. Then, the MGC of d wind turbines can be expressed as:

$$M(u_1, \ldots, u_d) = \int_0^{u_1} \cdots \int_0^{u_d} \frac{1}{(2\pi)^{d/2}|A|^{1/2}} \exp\left\{-\frac{1}{2}(u_1, \ldots, u_d)^T A^{-1} (u_1, \ldots, u_d)\right\} du_1 \ldots du_d \tag{3.38}$$

To sample wind outputs across different turbines, the Sobol sequence is employed. As a sampling method similar to quasi-Monte Carlo, Sobol sequence is more uniformly distributed compared to traditional sampling techniques [48, 49]. This uniformity enhances the accuracy of simulations.

The specific steps for implementing the Sobol sequence in modeling the coupling of uncertain wind outputs are detailed as bellow [50].

Step 1 Calculate the power output of wind turbine, if $v_{ci} < v_i < v_N$, $w_i = 0.5\pi\rho C_p(\lambda, \beta)R^2 v_i^3$; else if, $v_N < v_i < v_{co}$, $w_i = P_N$; otherwise, $w_i = 0$, where v_{ci}, v_{co} and v_N denote cut-in, cut-out, and nominal wind speed, respectively, P_N represents the wind turbine rated power, ρ denotes the air density (typically 1.25 km/m^3), λ refers to the pitch angle, R represents the blade radius (m), $C_p(\lambda, \beta)$ stands for the wind output coefficient.
Step 2 Obtain d wind turbines $[w_1, \ldots, w_i, \ldots, w_d]$ by using Sobol sequence.
Step 3 Generate $[u_1, \ldots, u_d]$, and then calculate the Sobol samples by the inverse conditional CDF of the MGC, i.e., $u_1 = w_1, u_2 = M_{2|1}^{-1}(w_2|u_1), \ldots, u_d = M_{d|1,\ldots,d-1}^{-1}(w_d|u_1, \ldots, u_{d-1})$, where $M_{i|1,\ldots,i-1}^{-1}$ represents the inverse Copula CDF.
Step 4 Obtain the spatially correlated samples of wind power from d different wind turbines $\left[P_1^W, \ldots, P_i^W, \ldots, P_d^W\right]$ by (3.39), where P_i^W denotes the power output of ith wind turbine.

$$P_i^W = F^{-1}(u_i) \tag{3.39}$$

This chapter seeks to minimize operational costs while adhering to risk constraints. The MSE model developed for the optimal operation of a low-carbon IES considering uncertain coupling is expressed as:

$$\begin{aligned}
&\min E[p_1 OC_1 + \ldots + p_i OC_i + \ldots + p_N OC_N] \\
&\text{s.t. } S_h[p_1 OC_1 + \ldots + p_i OC_i + \ldots + p_N OC_N] \le \gamma \\
&p_1 + p_2 + \ldots + p_N = 1 \\
&p_i \le 0, \ i = 1, 2, \ldots, N \\
&g(X, U) = 0 \\
&h(X, U) > 0
\end{aligned} \tag{3.40}$$

where E denotes the expectation cost of IES operation, OC_i represents the operation cost associated with the ith wind power sampling, p_i is the probability of ith sampling within N samples, γ signifies the risk level that IES operation need to tolerate, g denotes equality constraints, which includes (3.1)–(3.2), (3.6)–(3.7), (3.13), (3.17), (3.20)–(3.21) and (3.23)–(3.28), h refers to inequality constraints, including (3.4)–(3.5), (3.14)–(3.16), (3.18)–(3.19), (3.22) and (3.34). X is the vector of state variables, $X = \left[Im_1, \cdots, Im_{N_{Im}}, R_{U_1}, \cdots, R_{U_{N_U}}, R_{D_1}, \cdots, R_{D_{N_D}}, T_1^{ms}, \cdots, T_{N_{Ts}}^{ms}, T_1^{mr}, \cdots, T_{Nr}^{mr}\right]$, U indicates the vector related to control variables, $U = \left[P_1^C, \ldots, P_{N_C}^C, P_1^G, \ldots P_{N_G}^G, P_1^W, \ldots, P_{N_W}^W, P_1^{CHP}, \ldots, P_{N_{EB}}^{EB}\right.$ $\left.P_1^{EB}, \ldots, P_{N_{EB}}^{EB}, G_1^{GW}, \ldots, G_{N_{GW}}^{GW}\right]$.

To facilitate the calculation of the operation cost presented in Eq. (3.40), the procedures for fuzzy simulation are summarized in Algorithm 1.

Algorithm 1: Fuzzy simulation for operation cost

Set parameters $e = 0$, $i = 0$.
Obtain N vectors $x_1,\ldots, x_i,\ldots, x_N$ and then calculate the corresponding credibilities $\nu_1,\ldots, \nu_i, \ldots, \nu_N$ with $\nu_i = \nu(x_i)$, where the vector x_i denotes the sample of wind power for i sample.
Set two numbers
$OC_{\min} = OC_1(p_1, x_1) \wedge \ldots \wedge OC_N(p_N, x_N)$,
$OC_{\max} = OC_1(p_1, x_1) \vee \ldots \vee OC_N(p_N, x_N)$.
Randomly generate a real number r from $[OC_{\min}, OC_{\max}]$.
while $i \le N$ **do**
  if $r \ge 0$ **then**
    set $e \rightarrow e + Cr\{OC_i(p_i, x_i) \ge r\}$;
  else
    set $e \rightarrow e - Cr\{OC_i(p_i, x_i) \ge r\}$;
  end if
  $i = i + 1$;
end while
$E = OC_{\min} \vee 0 + OC_{\max} \wedge 0 + e \times (OC_{\max} - OC_{\min})/N$.

3.5 Case Study

The proposed MSE is evaluated using an IES that comprises the IEEE 39-bus power system, a regional 6-node heating system, and the Belgian 20-node natural gas system. Figure 3.4 depicts the configuration of the IES.

To derive the Mixed-Integer Linear Programming (MILP) model, linearization is performed using methods described in [51, 52], similar to the linearization approach for natural gas flow outlined in Sect. 2.5. The MILP model is then solved using Gurobi. In this setup, buses 33 and 37 in power system are connected to nodes 6 and 19 of natural gas system via gas-fired generators. Wind turbines are situated to

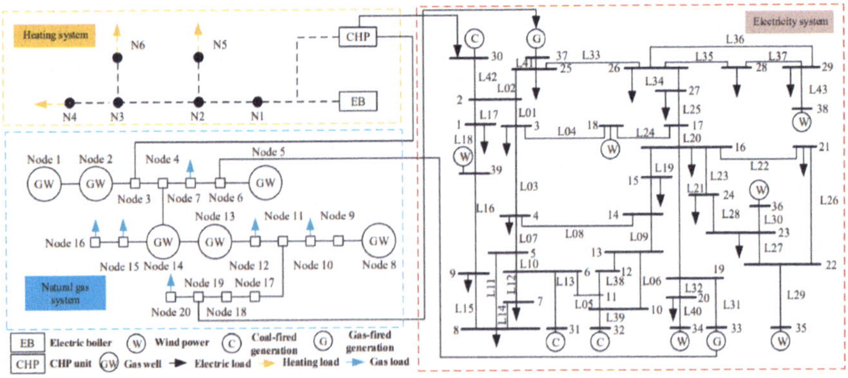

Fig. 3.4 Diagram of IES, reprinted from Ref. [57]. Copyright 2024, with permission Energy

buses 18, 34, 35, 36, 38, and 39, with each area having a wind power plant capacity of 6 MW, consisting of six turbines. As for the CHP and EB in heating system, they are linked to bus 30 of power system, with the CHP receiving gas from node 3 in the natural gas system.

Wind speed data for the turbines are sourced from [50], with data collected every 15 min. Forecasts for the next 15 min are based on 3000 historical wind speed records, with a sample size of 150 used for forecasting errors. Wind speed samples are derived by adding error samples to the forecasted wind speeds [50]. Each wind turbine has a rated power of 5 MW, with rated, cut-in, and cut-out wind speeds set at 12 m/s, 2 m/s, and 18 m/s, respectively [50].

The gas production costs are specified as \$0.085/m^3 for 13 gas wells and \$0.062/m^3 for gas wells numbered 4 through 6 [53]. The emission coefficient is 0.9 tons/MWh, with a emission cost of \$13.57 per ton [53]. For CHP, the fuel consumption coefficients are 2.40 kcf/MW for power production and 0.31 kcf/MW for heat production [53]. Both the electric boiler (EB) and CHP units have a power-heat conversion efficiency of 0.8. The carbon emission intensities are 1 ton/MW for coal-fired generators and 0.5 ton/MW for gas-fired generators. The carbon capture consumption is 0.2 MW/ton, and the carbon transmission and storage cost is \$3.19/ton [54].

Additionally, the flowchart illustrating the MSE model for low-carbon IES operation considering uncertain couplings is shown in Fig. 3.5.

3.5.1 Impact of Mean-Semi-entropy

To evaluate the impact of the MSE model on the IES, simulations are conducted with and without this model. The simulation without MSE is referred to as "Case 1", while the simulation incorporating MSE is labeled "Case 2". Detailed results for both cases are presented and compared in Table 3.1.

Table 3.1 presents the operation costs, emission costs, coal-fired generation, gas-fired generation, CHP generation, wind power generation, and electric boiler generation for both Case 1 and Case 2. The comparisons reveal that incorporating the MSE model in Case 2 results in a 2.29% reduction in total cost and a significant 44.43% decrease in emission cost compared to Case 1. Conversely, the penetration level of wind power increases in Case 2 relative to Case 1. Besides, the optimal hourly outputs of different generators are given by Fig. 3.6.

To further assess the proposed MSE, the comparison with two other models: the credibility-based fuzzy model and the probability-based mean–variance model [16]. The cases using these models are respectively referred to as "Case 3" and "Case 4". The comparisons are illustrated in Fig. 3.7.

The comparison reveals that, in both Case 3 and Case 4, the outputs of coal-fired and gas-fired generators increase while wind output decreases relative to Case 2. Specifically, the total outputs of coal-fired and gas-fired generators in Case 2 are 422.7477 MW, lower than the 445.4418 MW in Case 3 and 431.7674 MW in Case 4. Conversely, the total wind output in Case 2 is 659.6465 MW, exceeding the

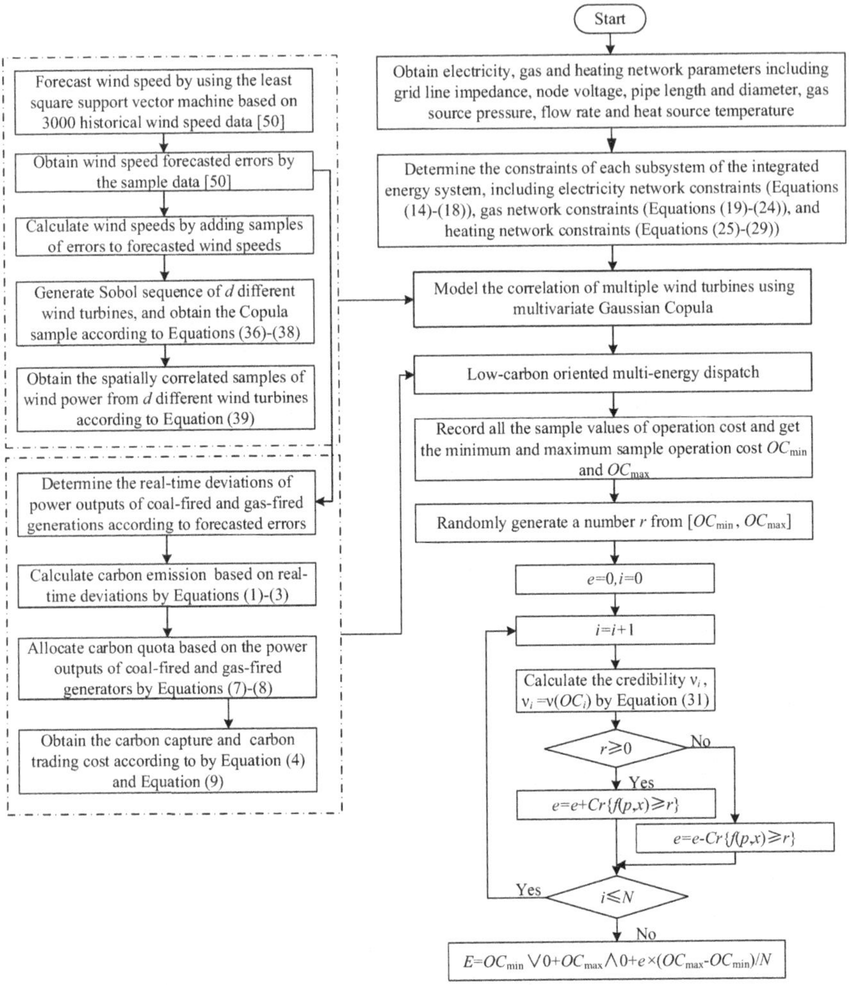

Fig. 3.5 Flowchart of proposed MSE model for IES

Table 3.1 Comparisons of dispatching results of Case 1 and Case 2, reprinted from Ref. [57]. Copyright 2024, with permission Energy

	Case 1	Case 2
Total cost (10^6\$)	4.0361	**3.9437**
Emission cost (10^5\$)	2.5946	**1.4419**
Coal-fired generation (MWh)	159.4882	**154.8196**
Gas-fired generation (MWh)	275.3335	**267.9281**
Wind power generation (MWh)	644.5865	**659.6465**
CHP generation (MWh)	36.8522	**43.3393**
Electric boiler generation (MWh)	15.9253	**6.4522**

The bold indicates better results.

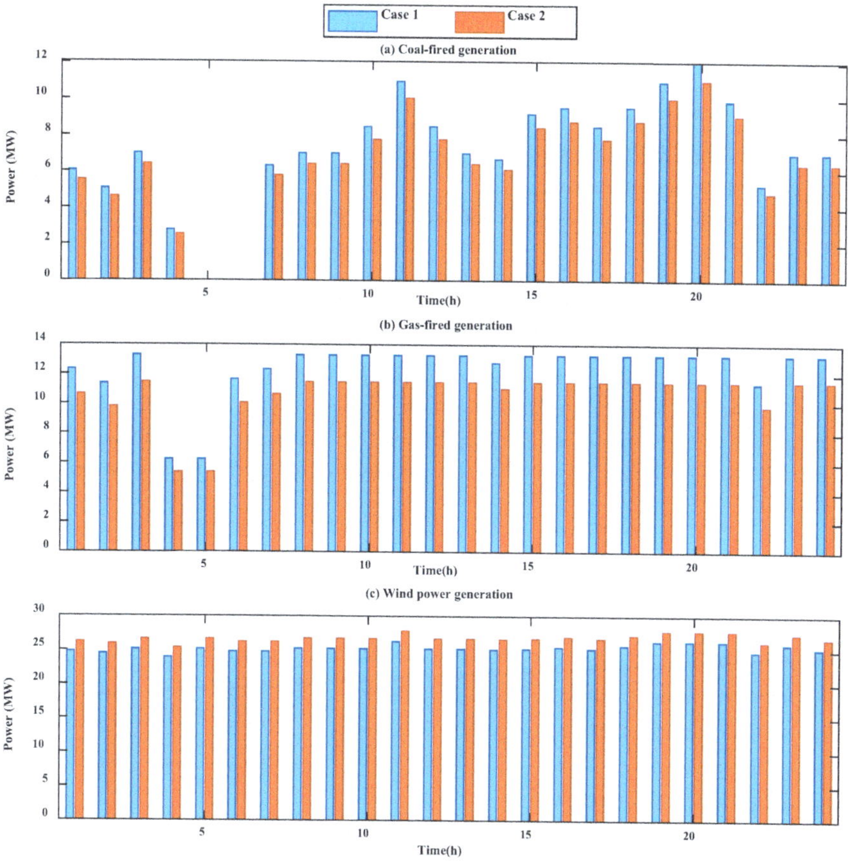

Fig. 3.6 Hourly scheduled outputs of coal-fired generation, gas-fired generation, and wind power

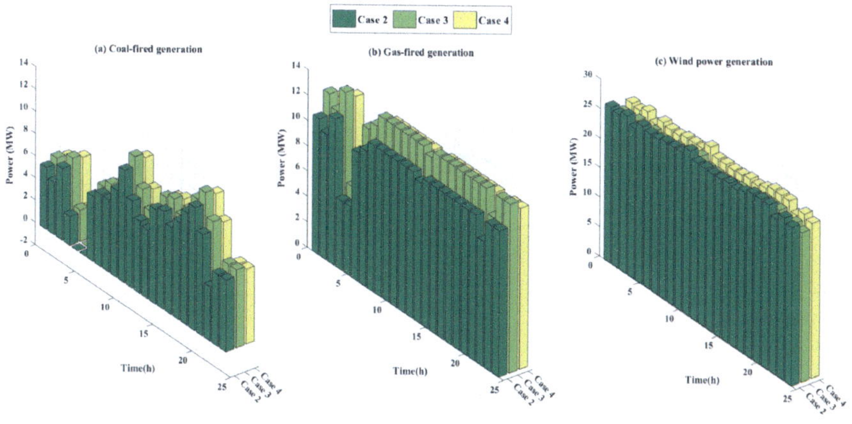

Fig. 3.7 Comparisons of MSE, credibility model, and mean–variance model

636.9472 MW in Case 3 and the 650.6268 MW in Case 4. Notably, while the wind output is higher in Case 3 compared to Case 2, it is still lower than that in Case 2.

As detailed in Table 3.2, the emission cost in Case 2 is reduced by 5.925×10^4\$ and 5.444×10^4\$ significant margins compared to both Case 3 and Case 4. This demonstrates that the proposed MSE model effectively captures the random and fuzzy characteristics in dealing with uncertain wind power. It provides a balanced approach to managing the trade-off between economic efficiency and uncertain risk in IES operation.

Additionally, Figs. 3.7 and 3.8 illustrate the hourly outputs for coal-fired generation, wind power and gas-fired generation over 24 h. These figures demonstrate that the proposed MSE effectively manages both operational costs and wind power integration within a given risk measure, outperforming the credibility model (Case 3) and the mean–variance model (Case 4). Consequently, it can be concluded that the proposed model offers a valuable approach for evaluating uncertain wind power in IES, providing a balanced solution that addresses both operational cost and risk.

Table 3.2 Comparisons of dispatching results of Case 2, Case 3, and Case 4, reprinted from Ref. [55]. Copyright 2024, with permission Energy

	Case 2	Case 3	Case 4
Total cost (10^6\$)	3.9437	4.0107	3.9822
Low-carbon emission cost (10^5\$)	1.4419	2.0344	1.9863
Total coal-fired generation (MWh)	154.8196	162.1083	156.0980
Total gas-fired generation (MWh)	267.9281	283.3335	275.6694
Total wind power generation (MWh)	659.6465	636.9472	650.6268
Total CHP generation (MWh)	43.3393	38.2231	39.6641
Total electric boiler generation (MWh)	6.4522	11.5736	10.1274

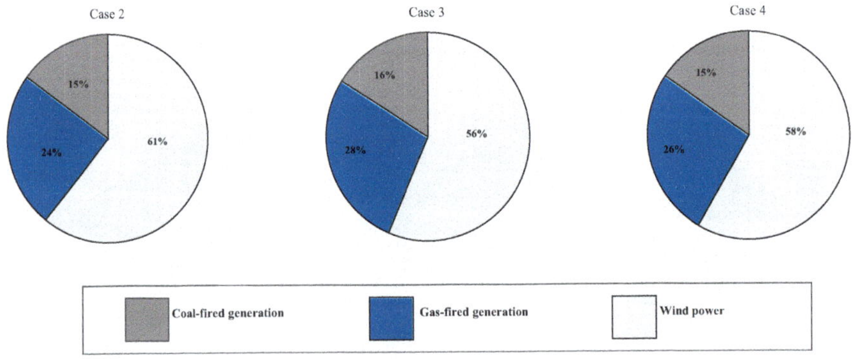

Fig. 3.8 Contribution of power supplied by coal-fired generation, gas-fired generation, and wind power

3.5.2 Impact of Low-Carbon Emission

To assess the impacts of incorporating low-carbon technologies in IES, real-time deviations of emissions are studied. The scenarios were categorized as follows: "Case 1" for the baseline without low-carbon technologies, "Case 5" for the scenario with low-carbon emission technologies, and "Case 6" for the scenario considering real-time deviations in carbon emissions due to uncertain wind power.

The results of interest, including operation cost, emission cost, and different generators outputs, are summarized and compared for Case 1, Case 5, and Case 6 in Table 3.3. The analysis indicates that integrating low-carbon technologies in Case 6 leads to a significant 5.78% reduction with respect to the total operational cost compared to Case 1, along with a substantial 63.63% decrease in emission costs. Notably, Case 6 shows a decrease of 32.3837 MWh by combining coal and natural gas generation and a 35.1688 MWh increase in total wind power generation relative to Case 1. Additionally, when compared to Case 5, Case 6 demonstrates an extra reduction in costs and an increase of 8.3188 MWh in wind power output. These findings are graphically depicted in Fig. 3.9, highlighting the significant of IES operations with low-carbon considerations.

The results underscore that adopting low-carbon technologies in IES not only offers economic benefits but also provides significant emission reduction. This approach, by balancing economic and environmental factors, represents a substantial improvement over traditional operational strategies.

As a whole, the comparisons shown in Table 2.4 and Fig. 3.9 highlight the critical role of low-carbon technologies in shaping the operation of IES thereby facilitating a transition toward a more environmentally and sustainable conscious landscape.

Furthermore, to assess the impacts of deviations in carbon emissions, a comparative analysis of the operation for Case 5 and Case 6 is depicted in Fig. 3.9. It is evident from this figure that accounting for real-time emission deviations leads to a significant reduction in outputs of both coal-fired and gas-fired generators, while simultaneously increasing outputs from wind turbines. This shift results in a corresponding

Table 3.3 Comparisons with and without low-carbon emission, reprinted from Ref. [55]. Copyright 2024, with permission Energy

	Case 1	Case 5	Case 6
Total cost (10^6\$)	4.0361	3.8271	**3.8029**
Emission cost (10^5\$)	2.5946	0.9623	**0.9437**
Coal-fired generation (MWh)	159.4882	139.4912	**136.0909**
Gas-fired generation (MWh)	275.3335	269.4666	**266.3471**
Wind power generation (MWh)	644.5865	671.4365	**679.7553**
CHP generation (MWh)	36.8522	42.4061	**41.5734**
Electric boiler generation (MWh)	15.9253	9.3854	**8.4190**

The bold indicates better results.

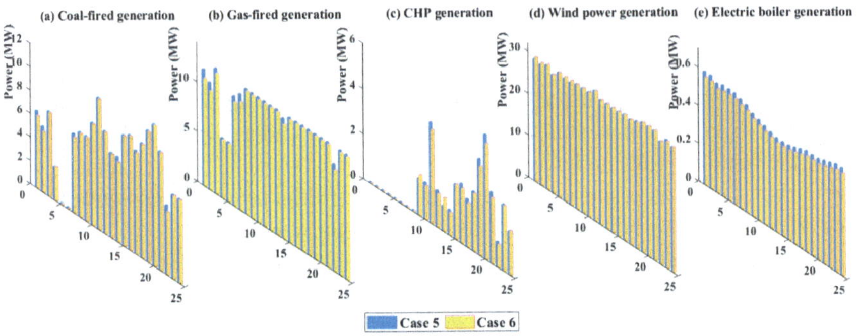

Fig. 3.9 Comparisons of optimal scheduling of Case 5 and Case 6. **a** Power generation of coal-fired generators. **b** Power generation of gas-fired generators. **c** Power generation of CHP generators. **d** Power generation of wind turbines. **e** Power consumption of EB

decrease in total carbon emissions. Specifically, the power outputs of gas-fired CHP and wind turbines are higher in Case 6 compared to Case 5, reflecting the beneficial effect of incorporating real-time emission deviations and low-carbon technologies. Thus, enhancing the share of gas-fired CHP and wind turbines with lower-emission coefficients proves to be a feasible strategy for reducing overall carbon emissions. Ultimately, maximizing power output from these lower-emission sources, rather than from coal-fired and gas-fired generators, performs more advantageous for the optimal operation of the integrated energy system.

3.5.3 *Impact of Multivariate Gaussian Copula*

The comparison of IES with and without considering uncertain wind outputs coupling is investigated in this subsection. Specifically, 6 wind turbines are located at node 18 as an example, as illustrated in Fig. 3.10. The scenarios with and without uncertain power outputs coupling of wind turbines are designated as "Case 1" and "Case 7", respectively. The optimization results for these two cases are compared and summarized in Table 3.4.

Combining the information from Fig. 3.10 and Table 3.4, it is evident that, in Case 7, where uncertain couplings are considered, the outputs of wind turbines are more compact compared to Case 1, where the power curves are more dispersed. Additionally, in Case 7, the outputs of wind turbines increases by 2.17%, and the operational cost decreases by \$2.352 $\times 10^5$ relative to Case 1. Thus, Case 7, which incorporates uncertain wind power couplings, enhances the correlation among different wind turbines, results in a more compact power distribution, and contributes to reduced fossil energy consumption while maintaining operational cost efficiency.

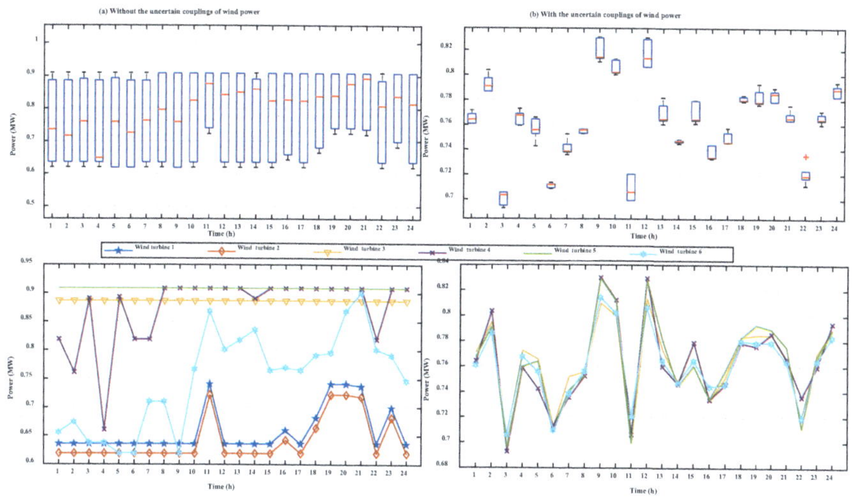

Fig. 3.10 Comparison of with and without couplings of wind turbines at bus 18

Table 3.4 Dispatching result with and without considering the uncertain couplings of wind turbines, reprinted from Ref. [55]. Copyright 2024, with permission Energy

	Case 1	Case 7
Total cost (10^6\$)	4.0361	**3.8009**
Emission cost(10^5\$)	2.5946	**0.8177**
Coal-fired generation (MWh)	159.4882	**142.4909**
Gas-fired generation (MWh)	275.3335	**281.3593**
Wind power generation (MWh)	644.5865	**658.5440**
CHP generation (MWh)	36.8522	**42.5228**
Electric boiler generation (MWh)	15.9253	**7.2687**

The bold indicates better results.

Additionally, the Kendall coefficient $\rho_{\rm k}$ and Spearman coefficient ρ_s are utilized to assess the linear and consistent correlations. Calculations reveal that ρ_s and ρ_k for Case 6 are 0.8696 and 0.9637, respectively, which are higher than the 0.7841 and 0.8945 observed in Case 1. This indicates a greater likelihood of similar variation trends in Case 6 compared to Case 1.

To further validate the effectiveness of multivariate Gaussian coupling in addressing the uncertain wind power, data from 36 wind turbines across six areas were used [50]. As illustrated in Fig. 3.11, the coupling model effectively captures the correlation of uncertain wind power within the same area. Particularly, for cases with high, low, or significant variations in wind power outputs, the model provides a more accurate depiction of the dependence structure among multiple wind turbines. This analysis aligns with actual data and supports the observed results.

Moreover, the data analysis in Fig. 3.11 corresponds to the optimization results from Case 6 shown in Fig. 3.10, where wind power outputs exhibit a more compact

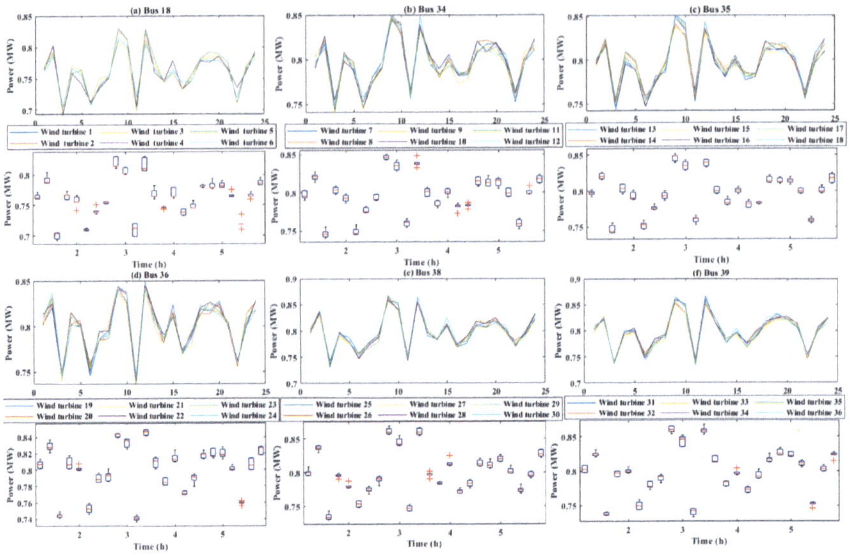

Fig. 3.11 Multivariate Gaussian couplings in six areas

distribution rather than a loose spread. Thus, incorporating the coupling model to study the uncertain outputs of wind turbines in IES operations offers a more realistic representation of wind turbine correlations, thereby minimizing errors associated with wind turbine output correlations.

The proposed model does not comprehensively address the coordination of wind power and energy storage, and missing a thorough exploration of the implications associated with integrating electric energy storage systems. Additionally, the model does not adequately incorporate several crucial aspects of flexibility. The potential benefits of these flexibility measures highlight the relative limitations of the current model. Addressing these aspects would significantly enhance the adaptability and robustness, particularly in managing the challenges presented by uncertain renewable energy.

3.6 Conclusion

This chapter proposes a MSE for low-carbon IES that encompass electricity, heating, and gas systems, accounting for the coupling of uncertain wind power. MSE incorporates low-carbon and addresses carbon emission deviations caused by uncertain wind power. By deriving the MSE, the downside risks referring to uncertain wind power can be effectively mitigated. Additionally, we utilize the multivariate Gaussian Copula to reveal the dependence structure among multiple wind turbines.

The MSE is rigorously tested on the IEEE 39-bus electricity system, a 6-node heating system and the Belgian 20-node natural gas system. Simulation results validate the model's effectiveness in managing IES operations to reduce both operation cost and carbon emission under uncertain wind power integration. Comparative analyses show that the proposed MSE approach surpasses other uncertain optimization models. The MSE model can be effectively used to quantify the fuzzy and stochastic characteristics of uncertain wind power, striking a balance between economic operation and risk aversion. Specifically, it reduces operational costs by 5.18% and contributes to a reduction in emissions, while optimizing outputs to enhance wind power integration.

Additionally, the multivariate Gaussian Copula proves to be a robust approach for capturing the complex dependencies and correlations among outputs from uncertain wind turbines. This results in a 2.17% increasing in wind power accommodation compared to scenarios without coupling, highlighting the importance of accounting for the interrelated nature of wind power rather than treating them as independent.

References

1. Zheng J, Guo J, Li Z, Wu Q, Zhou X (2023) Optimal design for a multi-level energy exploitation unit based on hydrogen storage combining methane reactor and carbon capture, utilization and storage. J Energy Storage 62:1–12
2. Liu J, Chen J, Yan G, Chen W, Xu B (2023) Clustering and dynamic recognition based auto-reservoir neural network: a wait-and-see approach for short-term park power load forecasting. iScience 26(8):1–23
3. Wan Y, Kober T, Schildhauer T, Schmidt TJ, McKenna R, Densing M (2023) Conditions for profitable operation of p2x energy hubs to meet local demand with energy market access. Adv Appl Energy 10:1–16
4. Ge L, Liu H, Yan J, Zhu X, Zhang S, Li Y (2022) Optimal integrated energy system planning with DG uncertainty affine model and carbon emissions charges. IEEE Trans Sustain Energy 13(2):905–918
5. Fu X, Zhou Y (2022) Collaborative optimization of pv greenhouses and clean energy systems in rural areas. IEEE Trans Sustain Energy 14(1):642–656
6. Zhang X, Bai Y, Zhang Y (2022) Collaborative optimization for a multi-energy system considering car- bon capture system and power to gas technology. Sustain Energy Technol Assess 49(101765):1–13
7. Zhang X, Zhang Y (2020) Environment-friendly and economical scheduling optimization for integrated energy system considering power-to-gas technology and carbon capture power plant. J Clean Prod 276(123348):1–16
8. Wang L, Shi Z, Dai W, Zhu L, Wang X, Cong H et al (2022) Two-stage stochastic planning for integrated energy systems accounting for carbon trading price uncertainty. Int J Electr Power Energy Syst 143(108452):1–20
9. Li Y, Bu F, Gao J, Li G (2022) Optimal dispatch of low-carbon integrated energy system considering nuclear heating and carbon trading. J Clean Prod 378(134540):1–15
10. Chen J, Qi B, Peng K, Li Y, Zhao Y (2020) Conditional value-at-credibility for random fuzzy wind power in demand response integrated multi-period economic emission dispatch. Appl Energy 261(114337):1–14
11. Gao Y, Wang J, Gao S, Cheng Y (2021) An integrated robust design and robust control strategy using the genetic algorithm. IEEE Trans Ind Inform 17(12):8378–8386

12. Siqin Z, Niu D, Li M, Gao T, Lu Y, Xu X (2022) Distributionally robust dispatching of multi-community integrated energy system considering energy sharing and profit allocation. Appl Energy 321(119202):1–15
13. Lyu X, Liu T, Liu X, He C, Nan L, Zeng H (2022) Low-carbon robust economic dispatch of park-level integrated energy system considering price-based demand response and vehicle-to-grid. Energy 213(125739):1–13
14. Markowitz HM (1968) Portf Sel 1:22–47
15. Xu D, Bai Z, Jin X, Yang X, Chen S, Zhou M (2022) A mean-variance portfolio optimization approach for high-renewable energy hub. Appl Energy 325:1–12
16. Li Y, Wu Q, Li M, Zhan J (2014) Mean-variance model for power system economic dispatch with wind power integrated. Energy 72(1):510–520
17. Chen J, Wu Q, Zhang L, Wu P (2017) Multi-objective mean–variance–skewness model for nonconvex and stochastic optimal power flow considering wind power and load uncertainties. Eur J Oper Res 263(2):719–732
18. Li X, Uysal AS, Mulvey JM (2022) Multi-period portfolio optimization using model predictive control with mean-variance and risk parity frameworks. Eur J Oper Res 299(3):1158–1176
19. Tan C, Wang J, Geng S, Pu L, Tan Z (2021) Three-level market optimization model of virtual power plant with carbon capture equipment considering copulacvar theory. Energy 237(121620):1–18
20. Mu Y, Wang C, Cao Y, Jia H, Zhang Q, Yu X (2022) A cvar-based risk assessment method for park- level integrated energy system considering the uncertainties and correlation of energy prices. Energy 247(123549):1–16
21. Kim D, Oh S (2022) Efficient off-policy safe reinforcement learning using trust region conditional value at risk. IEEE Robot Autom Lett 7(3):7644–7651
22. Gao J, Chen J, Qi B, Zhao Y, Peng K, Zhang X (2021) A cost-effective two-stage optimization model for microgrid planning and scheduling with compressed air energy storage and preventive maintenance. Int J Electr Power Energy Syst 125(106547):1–14
23. Lacagnina V, Pecorella A (2006) A stochastic soft constraints fuzzy model for a portfolio selection problem. Fuzzy Sets Syst 157(10):1317–1327
24. Wang B, Wang S, Zhou XZ, Watada J (2015) Two-stage multi-objective unit commitment optimization under hybrid uncertainties. IEEE Trans Power Syst 31(3):2266–2277
25. Jiang P, Yang H, Heng J (2019) A hybrid forecasting system based on fuzzy time series and multi-objective optimization for wind speed forecasting. Appl Energy 235:786–801
26. Miranda V, Hang P (2005) Economic dispatch model with fuzzy wind constraints and attitudes of dispatchers. IEEE Trans Power System 20(4):2143–2145
27. Mohammadi M, Noorollahi Y, Mohammadi-ivatloo B (2020) Fuzzy-based scheduling of wind integrated multi-energy systems under multiple uncertainties. Sustain Energy Technol Assess 37(100602):1–13
28. Patwal RS, Narang N (2020) Multi-objective generation scheduling of integrated energy system using fuzzy based surrogate worth trade-off approach. Renew Energy 156:864–882
29. Liu D et al (2007) Uncertainty theory. Springer
30. Baoding L, Jin P (2005) A course in uncertainty theory. Eur J Oper Res 7:74–126
31. Wang B, Wang S, Zhou X, Watada J (2016) Multi-objective unit commitment with wind penetration and emission concerns under stochastic and fuzzy uncertainties. Energy 111:18–31
32. Zhou J, Li X, Pedrycz W (2016) Mean-semi-entropy models of fuzzy portfolio selection. IEEE Trans Fuzzy Syst 24(6):1627–1636
33. Huang X (2008) Mean-entropy models for fuzzy portfolio selection. IEEE Trans Fuzzy Syst 16(4):1096–1101
34. Guo Y, Ming B, Huang Q, Liu P, Wang Y, Fang W et al (2022) Evaluating effects of battery storage on day-ahead generation scheduling of large hydro wind photovoltaic complementary systems. Appl Energy 324:1–21
35. Wald D, Johnson K, King J, Comden J, Bay CJ, Chintala R et al (2023) Shifting demand: reduction in necessary storage capacity through tracking of renewable energy generation. Adv Appl Energy 10(100131):1–17

36. Quan H, Lv J, Guo J, Zhang W (2022) Investigation of spatial correlation on optimal power flow with high penetration of wind power: a comparative study. Appl Energy 316(119034):1–10
37. Xu X, Yan Z (2017) Probabilistic load flow calculation with quasi-monte Carlo and multiple linear regression. Int J Electr Power Energy Syst 88:1–12
38. Hajian M, Rosehart WD, Zareipour H (2012) Probabilistic power flow by monte Carlo simulation with Latin supercube sampling. IEEE Trans Power Syst 28(2):1550–1559
39. Xiao Q (2015) Comparing three methods for solving probabilistic optimal power flow. Electr Power Syst Res 124:92–99
40. Jiao P, Chen J, Cai X, Wang L, Zhao Y, Zhang X et al (2021) Joint active and reactive for allocation of renewable energy and energy storage under uncertain coupling. Appl Energy 302(117582):1–19
41. Mandelbrot BB (1997) The variation of certain speculative prices, vol. 7, pp 371–418
42. Villanueva D, Feij´oo A, Pazos JL (2013) Multivariate weibull distribution for wind speed and wind power behavior assessment. Resources 2(3):370–384
43. Wang Z, Shen C, Liu F, Wu X, Liu CC, Gao F (2017) Chance-constrained economic dispatch with non-gaussian correlated wind power uncertainty. IEEE Trans Power Syst 32(6):4880–4893
44. Ding T, Xu Y, Wei W, Wu L (2019) Energy flow optimization for integrated power–gas generation and transmission systems. IEEE Trans Ind Inf 16(3):1677–1687
45. Zadeh LA (1978) Fuzzy sets as a basis for a theory of possibility. Fuzzy Sets Syst 1(1):3–28
46. Zadeh LA (1979) A theory of approximate reasoning. Mach Intell 9(9):149–194
47. Scott DW (2015) Multivariate density estimation: theory, practice, and visualization. John Wiley & Sons
48. Niederreiter H (1978) Quasi-monte Carlo methods and pseudo-random numbers. Bull Am Math Soc 84(6):957–1041
49. Niederreiter H (1992) Random number generation and quasi-Monte Carlo methods. SIAM
50. Xie Z, Ji T, Li M, Wu Q (2017) Quasi-monte Carlo based probabilistic optimal power flow considering the correlation of wind speeds using copula function. IEEE Trans Power Syst 33(2):2239–2247
51. Liu H, Shen X, Guo Q, Sun H (2021) A data-driven approach towards fast economic dispatch in electricity–gas coupled systems based on artificial neural network. Appl Energy 286:1–18
52. Vandewalle J (1975) On the calculation of the piecewise linear approximation to a discrete function. IEEE Trans Comput 100(8):843–846
53. Jiang T, Deng H, Bai L, Zhang R, Li X, Chen H (2018) Optimal energy flow and nodal energy pricing in carbon emission-embedded integrated energy systems. CSEE J Power Energy Syst 4(2):179–187
54. Jiao P, Cai X, Wang L, Chen J, Zhao Y, Cao Y (2023) Flexibility operation for integrated energy system considering hydrogen energy under inertia characteristics and stepped carbon trading mechanism. Sustain Cities Soc 98:1–18
55. Jiao PH, Chen JJ, Cai X et al (2024) Fuzzy semi-entropy based downside risk to low-carbon oriented multi-energy dispatch considering multiple dependent uncertainties. Energy 287:129717

Part II
Renewable Distribution Network and Microgrid Optimization

Chapter 4
Multi-step Reconfiguration with Many-Objective Reduction for Renewable Distribution System

4.1 Introduction

The large-scale integration of Distributed Renewable Energy (DRE) into distribution networks emerges as a prominent trend expected to persist in the foreseeable future [1, 2]. Unlike conventional centralized power plants, DRE facilities are situated closer to user side, generating electricity at lower voltage levels. However, the significant scale of DRE integration poses great operational challenges for distribution networks [3, 4]. To effectively manage these challenges, collaborative efforts between utilities and policymakers are essential to develop innovative technologies and strategies tailored to the integration of DRE systems within distribution networks.

Among the strategies considered, Distribution Network Reconfiguration (DNR) stands out as a cost-effective approach to mitigating power losses. DNR methodology does not necessitate the addition of external devices, thereby avoiding extra installation expenses. The distribution network comprises an array of segmented and contact switches, operating under closed-loop design and open-loop modes [5]. The fundamental principle of DNR involves adjusting the on/off status of sectional and contact switches to modify the network's topology, aiming to enhance system performance in terms of improving voltage quality and reducing power losses [6]. The concept of DNR was initially introduced in 1975 [7], and since then, researchers have extensively explored this area with the primary objective of minimizing power loss through network optimization.

In the pursuit of effective DNR, several methodologies have been explored in the literature. Singh et al. [8] introduce a sequential switch opening method aimed at minimizing power loss within the feeder configuration. This approach strategically incorporates branches with minimal resistivity currents to optimize network performance. Similarly, Ghasemi et al. [9] propose a branch switching technique to enhance DNR outcomes, focusing on both system efficiency and economic considerations.

J. Chen and Y. Li, *Renewable Power System Optimization*, Smart Energy Systems,
https://doi.org/10.1007/978-981-97-8132-4_4

While these individual techniques offer valuable insights, standalone implementations may not always yield the desired level of precision. To address this, Zhan et al. [10] combine sequential switch opening with branch switching, introducing a Switch Opening and Exchange (SOE). This multi-step reconfiguration strategy has been shown to outperform singular methodologies, providing enhanced accuracy in network optimization. In a related vein, Harsh et al. [11] present a two-stage heuristic approach grounded in sequential switch operations, designed to achieve rapid DNR. Despite the advancements made by Zhan et al. [10] and Harsh et al. [11], it is notable that their focus primarily centers on power loss minimization as the primary optimization objective. This leaves an avenue for further exploration into additional factors and objectives that could be considered within the context of DNR, such as voltage stability, load balancing, and economic efficiency.

Furthermore, the challenges in distribution network operation necessitate the consideration of additional flexibility indicators beyond power loss, factors that significantly impact both the economic efficiency and stability of the system. Given the inherent variability and randomness in DRE output, it becomes imperative to study how its widespread integration influences different flexibility indices within the distribution system. These indices include operational economy metrics such as operation cost and network losses, as well as aspects of system stability like load fluctuations, voltage quality, and peak-load management.

In a study by [12], the focus is on optimizing the reconfiguration of distribution systems using a multi-objective optimization. The primary objective is to enhance the overall stability of the distribution network while simultaneously minimizing network losses. Another approach, as described in [13], formulates a multi-objective DNR incorporating several optimization objectives, such as voltage stability and active power loss indices. The analytic hierarchy process is utilized to synthesize these multi-objective considerations, aiming for a comprehensive optimization strategy. Additionally, [14] proposes an advanced optimization algorithm based on Particle Swarm Optimization (PSO). This algorithm deals with the multi-objective DNR by simultaneously optimizing three distinct objectives: power loss reduction, enhancement of voltage stability indices, and minimizing switching operations. This method represents a sophisticated approach to solving the intricate optimization challenges inherent in DNR. Exploring further, [15] delves into the diverse topological configurations achievable through reconfiguration, trying to minimize the dispatching and control of flexible resources within the system. The study by [16] establishes and evaluates flexibility indices for power system operations, particularly considering the growing availability and integration of DRE within the grid.

In practice, the operation of distribution networks must contend with a multitude of flexibility indicators, presenting a challenge of many-objective and high-dimensional optimization. Traditional multi-objective algorithms, while effective for two- or three-objective scenarios, encounter significant limitations when confronted with many-objective and high-dimensional optimization problems. This limitation stems from the exponential degradation in performance as a result of the incompatibility of the majority of solutions within the Pareto dominance criterion [17, 18].

To mitigate this challenge, strategies have been studied to address the issue of many-objective computing. Reference [19], for instance, proposed a dimension reduction approach in the realm of many-objective optimization, aiming to identify and select the most conflicting objectives. Simulation results demonstrate that the dimension reduction approach significantly enhances the effectiveness in Pareto dominance criteria. However, this technique may prove futile in scenarios where no duplicate objectives exist, rendering the objective reduction strategy ineffective.

In an effort to refine objective selection, the concept of an aggregation tree has been explored in works such as [20, 21]. This method aims to guide the selection process toward the most harmonious alternative, yet it may not always yield the optimal objective configuration. Recognizing this limitation, an enhanced objective reduction strategy was introduced in [22], utilizing the Spearman coefficient to assess the interrelationship among different objectives.

The presence of DRE prediction errors significantly impacts the operational costs of the system and influences the optimal design of DNR, particularly due to the unreliable nature of day-ahead scheduling in forecasting DRE power generation [23, 24]. Additionally, [25] presents both hard and soft strategies to mitigate the impact of uncertainties from DRE on DNR. In the realm of risk-based DNR, [26] develops a model to account for the unpredictability in load and power generation. This study utilizes scenario theory and PSO to address the challenges associated with risk modeling and optimization. While [27] discusses strategies for integrating active and reactive power coupling, the focus does not extend to the technologies related to DNR.

To delineate the contributions of this study, we undertake a comparative analysis with related literature and find several key observations: (1) The prevailing research in distribution network scheduling with DRE predominantly centers on enhancing power loss reduction or improving voltage quality. However, the operational challenges of distribution networks extend beyond these parameters to encompass additional flexibility indicators impacting stability and economics. (2) Existing methodologies for DNR in stochastic distribution network scheduling largely rely on branch switching or sequential switch opening techniques. Yet, employing a singular technique may not yield adequately accurate results in DNR optimization. (3) Current stochastic scheduling approaches with DNR often treat uncertain DRE as a constraint, which may present limitations in evaluating system operational safety. Moreover, such methods might struggle to strike a balance between accommodating and discarding DRE resources.

Drawing from these insights, our study is guided by three primary research objectives: (1) Investigate the influence of widespread DRE availability on diverse flexibility indices in distribution system operation, encompassing factors such as operational costs, network losses, and stability metrics including load fluctuation and voltage quality. (2) Explore the synergistic potential of combining branch switching and sequential switch opening techniques to enhance operational flexibility in stochastic distribution networks. (3) Examine the risk associated with uncertain DRE in the context of DNR, aiming to optimize penetration levels of DRE considering risk aversion.

The contributions of this chapter are outlined as follows:

(1) An iterative bi-level scheduling approach that combines multi-step reconfiguration with many-objective reduction techniques is proposed. This scheduling framework comprehensively addresses economic and stability concerns, encompassing factors such as operation cost, power loss, voltage fluctuations, net load variability, and power load rates.
(2) At the upper level of our framework, a Multi-Step Switching Sequence Exchange (MSSE) approach is developed. It focuses on minimizing power loss through DNR. The MSSE method entails a sequential switch opening strategy for determining switch states, coupled with branch switching.
(3) Building upon the topology established in the upper level, an objective reduction is introduced at the lower level. This model aims to enhance the economic efficiency and operational stability of stochastic distribution networks. A joint probability distribution leveraging forecast error and coupling information is developed to address the uncertain risks.

The remainder of this chapter is structured as follows. In Sect. 4.2, the problem model of the stochastic DNR with MSSE is presented. Section 4.3 describes the multi-objective reduction optimization. Simulation results are presented in Sect. 4.4. A conclusion is reached in Sect. 4.5.

4.2 Distribution Reconfiguration with MEES

4.2.1 MSSE Integrated DNR Model

A MSSE with DNR is proposed to minimize power loss in the upper level. The predicted output of DRE serves as the input for the MESS. Then optimizing the network topology to achieve the power loss minimum [28]:

$$\min \sum_{t=1}^{T} \sum_{i=1,j=1}^{N_n} \beta_{ij}^t \left(\frac{P_{ij}^t + Q_{ij}^t}{V_i^t} \right)^2 R_{ij} \tag{4.1}$$

where P_{ij}^t and Q_{ij}^t denote the active power and reactive power flowing from nodes i to j at time t, V_i^t represents the node voltage of i at time t, and R_{ij} is the branch resistance associated with nodes i and j, T indicates the number of scheduling intervals, β_{ij}^t is the branch switching state with respect to nodes i and j at time t, where 1 indicates closure and 0 indicates disconnection.

The constraints governing power flow in the distribution network are described by several equality and inequality equations. These constraints for DNR not only consist of the network topology constraints but also adherence to the physical limits requirement of electrical equipment.

(1) Flow constraints:

$$
\begin{aligned}
&P_{\mathrm{DRE},i}^{t}+P_{\mathrm{g},i}^{t}+P_{\mathrm{EPNS},i}^{t}-P_{\mathrm{dem},i}^{t}=\sum_{i,j\in\Omega_B}\beta_{ij}^{t}V_i^tV_j^t\left(G_{ij}\cos\delta_{ij}^{t}+B_{ij}\sin\delta_{ij}\right)\\
&Q_{\mathrm{DRE},i}^{t}+Q_{\mathrm{g},i}^{t}+Q_{\mathrm{EPNS},i}^{t}-Q_{\mathrm{dem},i}^{t}=\sum_{i,j\in\Omega_B}\beta_{ij}^{t}V_i^tV_j^t\left(G_{ij}\cos\delta_{ij}^{t}+B_{ij}\sin\delta_{ij}^{t}\right)\\
&\beta_{ij}^{t}\left(V_i^tV_j^t\right)^2\left[\left(G_{ij}\cos\delta_{ij}^{t}+B_{ij}\sin\delta_{ij}^{t}\right)^2+\left(G_{ij}\sin\delta_{ij}^{t}-B_{ij}\cos\delta_{ij}^{t}\right)^2\right]\le\beta_{ij}^{t}\overline{S}_{ij}^{2}n
\end{aligned}
\tag{4.2}
$$

where $P_{\mathrm{DRE},i}^{t}$ and $Q_{\mathrm{DRE},i}^{t}$ are respectively the active and reactive power of DRE of node i during time t. $P_{g,i}^{t}$ and $Q_{g,i}^{t}$ denote active and reactive power of the micro-turbine injected into node i during time t, respectively. $P_{\mathrm{EPNS},i}^{t}$ and $Q_{\mathrm{EPNS},i}^{t}$ represent reserve power of active and reactive in node i during time t, respectively. $P_{\mathrm{dem},i}^{t}$ and $Q_{\mathrm{dem},i}^{t}$ indicate demand power of active and reactive power in node i during time t, respectively. V_j^t represents the voltage amplitude at nodes j, and δ_{ij}^{t} is difference of voltage phase angle between nodes i and j. G_{ij} and B_{ij} denote imaginary and real parts of the ith row and jth column of the admittance matrix, respectively, $\overline{S}_{ij}$ is the maximum allowed power capacity between nodes i and j.

(2) Voltage and current constraints:

$$
\begin{cases}
V_{i,\min}\le V_i^t\le V_{i,\max}\\
0\le I_{ij}^{t}\le I_{ij,\max}
\end{cases}
\tag{4.3}
$$

where $V_{i,\min}$ and $V_{i,\max}$ are the upper and lower voltage limits for node i, respectively, I_{ij}^{t} denotes the branch current between nodes i and j during time t, and $I_{ij,\max}$ denotes the maximum current.

(3) Micro-turbine power output constraints:

$$
\begin{cases}
0\le P_{g(m)}^{t}\le P_{g,\max}\\
Q_{g,\min}\le Q_{g(m)}^{t}\le Q_{g,\max}
\end{cases}
\tag{4.4}
$$

where $P_{g,\max}$ denotes the upper bound of micro-turbine active power output, $Q_{g,\max}$ and $Q_{g,\min}$ represent the maximum and minimum bounds of micro-turbine reactive power output, respectively, $Q^{t}_{g(m)}$ and $P^{t}_{g(m)}$ indicate the reactive and active power output of mth micro-turbine, respectively.

(4) Radiality and connection constraints:

DNR process must adhere to connectivity and radiality constraints in addition to the aforementioned constraints. It should be mentioned that the network must remain radial even after reconfiguration, with no isolated nodes. The system must ensure that the radiality and connectivity of the network are maintained, following these restrictions:

$$N\_\{n\} - N\_\{b\} = 1 \tag{4.5}$$

$$\mathrm{dis}(i, i_{\mathrm{sub}}) < \infty \quad (i \neq i_{\mathrm{sub}}) \tag{4.6}$$

where Eq. (4.5) denotes the relationship between nodes and closed branches, N_b represents the number of branches. Equations (4.5) and (4.6) ensure the connectivity and radiality of the network. Additionally, $\mathrm{dis}(i, i_{\mathrm{sub}})$ denotes the electrical distance from node i to the substation node i_{sub}.

4.2.2 *Multi-step Switching Sequence Exchange Method*

The proposed DNR, referred to as the MSSE approach, is divided into three steps. First, the sequential switching method is applied. This step involves analyzing and determining the optimal switch states for sequential switch openings by minimizing power loss. Second, the branch switching method is employed. This step focuses on adjusting the branch switches' states to further optimize the network. By modifying these branches states, it can reach a more efficient system operation in reducing power loss. These steps are critical in the MSSE approach to DNR.

In the first step, all switches are initially turned off to form a loop topology. The process involves determining the tie switch with the minimum power loss through sequential iterations. Specifically, each switch in the loop topology is sequentially opened, and the corresponding power loss is calculated. After completing this process for all switches, the switch with the lowest power loss is identified as the tie switch. Subsequently, all contact switches are turned on to achieve a radial topology. The pseudo-code for the first step of the MSSE method is described in Algorithm 1.

Algorithm 1: The pseudo code of first-step for MSSE

Input: Load parameters, initial topology, power generation, active and reactive power output of DRE and micro-turbine.
Output: optimal topology under first-step.

stop $= 0$
while stop $= 0$ **do**
  for each switch ι in loops **do**
    $\delta_\iota = 0$
    if dis$(i, i_{sub}) < \infty$ **then**
The DNR model is solved, and power loss obtained by (1) is saved in $L_1(\iota)$.
If any constraints are not satisfied, let $L_1(\iota) = 10^9$
    else
      $L_1(\iota) = 1 \times 10^9$
    end if
    $\delta_\iota = 1$
  end for
Find and turn on the switch with minimum loss. $k = \arg\min L_1(\iota)$, $\delta_k = 0$.
  if dis$(i, i_{sub}) >= \infty$ **then**
    stop $= 1$
  end if
end while

The first step generates a basis network topology for the second step. In the second step, the sectionalizing switch of the obtained topology in the first step is converted to the tie switch. To achieve a better solution, the power loss at the upper level under different topologies is investigated based on the power outputs of micro-turbine and DRE. Additionally, the switching states of three types of switches are kept unchanged to reduce the computational burden.

Type-1 switches: Any sectionalizing switch whose number of branches to the substation node is less than a predefined threshold n_1.

Type-2 switches: Any sectionalizing switch where the shortest path to the end node is less than a specified limit n_2.

Type-3 switches: Any sectionalizing switch that is not part of a switch closing loop.

Algorithm 2: The pseudo code of second and third steps for MSSE

Input: optimal topology of first step, n_1, n_2, all type-2 switches S_e
Output: optimum topology and minimal power loss
In the topology of upper level, the set S_ι is all sectionalizing switches
Exclude type-1, type-2, and type-3 switches from S_ι
for each switch ι in S_ι **do**
  Let $\delta_\iota = 0$ and $\delta_k = 1$. The first step is completed, and the topology obtained is saved. The objective function (1) is calculated and the resulting power loss is stored
  for each switch ι in S_e **do**
    Find all on–off actions that contain opening switch ι
    for each on–off action **do**
      Calculate objective function (1), denoted as l_0, save the result to $L_2(\iota)$ and save corresponding topology
      if $L_2(\iota) < L_1(\iota)$ **then**
        Store the on–off action in S_3
      end if
    end for
  end for
  Determine the best network topology from L_1 and L_2. $\delta_\iota = 1$
end for

In the initial topology, the state of type-1 switches generally does not require modification since switches close to the substation node are easy to identify. Additionally, opening type-3 switches is prohibited to prevent node isolation. Type-2 switches remain unmodified in the second phase due to a more efficient approach for state changes proposed in the third step.

In the third step, a new topology is generated by modifying the state of type-2 switches identified in the second step. This modification ensures network connectivity while power loss minimum. Specifically, the third step is an on–off action (i.e., opening type-2 switches) and closing a tie switch to maintain network connectivity. The DNR that results in the minimum power loss is selected. The pseudo-codes for the second and third steps of MSSE are detailed in Algorithm 2.

4.3 Many-Objective and Active and Reactive Power Coupling

To incorporate the stability and economy of DNR, in the lower level, this subsection develops a multi-objective reduction model, based on the topology optimized from the upper level. Besides, the prediction errors and coupling information are incorporated to create a joint distribution, thereby accounting for the uncertain risks associated with the active and reactive power of DRE sources.

4.3.1 Objective Functions

The objectives in the lower level include operation cost, power loss, voltage variation, net load factor, net load peak-valley difference and power load rate.

(1) Power loss:

The integration of DRE into the power network impacts the distribution and flow of power, typically increasing power loss. Therefore, minimizing power loss F_{loss} is an essential optimization objective to ensure the economic efficiency of the distribution network:

$$F_{\text{loss}} = \sum_{t=1}^{T} \sum_{i=1,j=1}^{N_n} \left(\frac{P_{ij}^t + Q_{ij}^t}{V_i^t} \right)^2 R_{ij} \tag{4.7}$$

(2) Operation cost:

The operational cost of the distribution network F_{cost} encompasses the generation cost of micro-turbines, electricity purchasing cost, and the DRE on-grid revenue. This is represented as follows:

$$F_{\text{cost}} = \sum_{t}^{T} \sum_{m}^{M} \left(a_m \left(P_{g(m)}^t \right)^2 + b_m P_{g(m)}^t + c_m \right) + \sum_{t}^{T} d^t P_L^t - \sum_{t=1}^{T} \rho P_{\text{EPC}}^t \tag{4.8}$$

where the first term of Eq. (4.8) denotes the micro-turbine generation cost. Specifically, $P_{g(m)}^t$ represents the power output of the mth micro-turbine at time t, a_m, b_m and c_m indicate the cost coefficients of the mth micro-turbine generation, and M is the number of micro-turbines. The second term of Eq. (4.8) reflects electricity cost. N_n represents the number of load nodes, d^t signifies time-of-use price, and P_{L}^t illustrates the electricity purchasing from upper grid at time t. The last term of Eq. (4.8) denotes on-grid revenue. ρ indicates the on-grid price, and P_{EPC}^t indicates electricity sold to the superior grid when the DRE output exceeds the predicted value at time t.

(3) Voltage deviation:

Voltage deviation is a critical indicator of power quality. High penetration of DRE can exacerbate voltage variations within the distribution network, potentially compromising system stability and adversely affecting the lifespan of equipment.

Equation (4.9) illustrates the average of node voltage deviation, with a rated value of V_{ref} [29].

$$F_{\text{VD}} = \frac{1}{N_n} \sum_{i=1}^{N_n} \sum_{t}^{T} \left| V_i^t - V_{\text{ref}} \right| \tag{4.9}$$

(4) Net load peak-valley difference:

Modifying the load peak-valley difference and smoothing the load curve can enhance the power system's ability to accommodate DRE. Consequently, the peak-valley difference of the net load is formulated as a flexibility indicator, as expressed in the following equation:

$$F_{p-v} = \frac{P_{L,\max} - P_{L,\min}}{P_{L,\max}} \tag{4.10}$$

where $P_{L,\max}$ and $P_{L,\min}$ represent the maximum and minimum net loads within T hours, respectively.

(5) Net load factor:

This study assesses the smoothness of the load curve by defining the load factor. A net load factor closer to 1 indicates a smoother net load curve. The relevant equation is as follows:

$$F_{\text{NL}} = \frac{\sum_{t=1}^{T} P_L^t}{P_{L,\max} T} \tag{4.11}$$

(6) Power load rate:

Currently, the cost of generating power with DRE installed on the user side is lower than purchasing electricity from the grid. Consequently, users are increasingly installing DRE systems to reduce their electricity cost. However, due to the variability and intermittency of DRE, the power grid must invest in additional equipments with significantly larger capacities, which results in decreased equipment utilization. To address this issue, it is essential to improve the load rate of the distribution network.

The power load rate is defined as the ratio of a user's maximum interactive power to the maximum load demand at a node within the system. The expression of the power load rate is provided below.

$$F_{\text{PL}} = \frac{\max\left\{P_{L(1)}^t, P_{L(2)}^t, \ldots, P_{L(m)}^t\right\}}{\max\left\{P_{\text{load}(1)}^t, P_{\text{load}(2)}^t, \ldots, P_{\text{load}(m)}^t\right\}} \quad m \in \Omega_M, t \in T \tag{4.12}$$

where Ω_M represents the set of load nodes with DRE, $P_{L(m)}^t$ refers to the interaction power between node m and the superior grid at time t, and $P_{\text{load}\,(m)}^t$ indicates load demand of node m at time t.

In view of the above discussion, the optimization objective is formulated as follows:

$$\min\left(F_{\text{cost}}, F_{\text{loss}}, F_{\text{VD}}, F_{\text{p-v}}, -F_{\text{NL}}, -F_{\text{PL}}\right) \tag{4.13}$$

4.3.2 Constraint Conditions

Besides the constraints (4.3) and (4.4) from the upper level, the lower level model must satisfy the following requirements.

(1) Flow constraints:

The network topology, derived from the DNR optimization at the upper level, undergoes multi-objective optimization at the lower level. Therefore, under the power flow limits specified at the lower level, it is not necessary to account for changes in branch switch states.

$$\begin{aligned} &P_{\mathrm{DRE},i}^{t} + P_{\mathrm{g},i}^{t} + P_{\mathrm{EPNS},i}^{t} - P_{\mathrm{dem},i}^{t} = \sum_{i,j\in\Omega_B} V_i^t V_j^t \left(G_{ij}\cos\delta_{ij}^t + B_{ij}\sin\delta_{ij}\right) \\ &Q_{\mathrm{DRE},i}^{t} + Q_{\mathrm{g},i}^{t} + Q_{\mathrm{EPNS},i}^{t} - Q_{\mathrm{dem},i}^{t} = \sum_{i,j\in\Omega_B} V_i^t V_j^t \left(G_{ij}\cos\delta_{ij}^t + B_{ij}\sin\delta_{ij}^t\right) \\ &\left(V_i^t V_j^t\right)^2\left[\left(G_{ij}\cos\delta_{ij}^t + B_{ij}\sin\delta_{ij}^t\right)^2 + \left(G_{ij}\sin\delta_{ij}^t - B_{ij}\cos\delta_{ij}^t\right)^2\right] \le \overline{S}_{ij}^2 \end{aligned} \tag{4.14}$$

(2) DRE power output constraints:

To avoid the circulation of reactive power in DNR, DRE units must operate with a sufficiently high power factor. Consequently, the operational constraints for DRE are described by the following expression:

$$\begin{cases} \dfrac{P_{\mathrm{DRE}(m)}^{t}}{\sqrt{\left(P_{\mathrm{DRE}(m)}^{t}\right)^2 + \left(Q_{\mathrm{DRE}(m)}^{t}\right)^2}} \ge f_{\mathrm{DRE}}^{t} \\ \left(P_{\mathrm{DRE}(m)}^{t}\right)^2 + \left(Q_{\mathrm{DRE}(m)}^{t}\right)^2 \le S_{\mathrm{DRE}(m)}^2 \end{cases} \tag{4.15}$$

where $P_{\mathrm{DRE}(m)}^{t}$ and $Q_{\mathrm{DRE}(m)}^{t}$ denote the active and reactive power output of the mth DRE at time t, respectively, f_{DRE}^{t} represents DRE's the minimum allowable power factor at time t, and $S_{\mathrm{DRE}(m)}$ refers to DRE's the apparent power.

4.3.3 Joint Probability-Based DRE Active and Reactive Power Coupling

Due to the inherent variability in DRE power output, accurately predicting their generation with high precision is challenging. Consequently, it is essential to consider the potential inaccuracies in power output forecasts when managing distribution networks that incorporate DRE sources. From a risk management perspective, the economic risk and economic loss associated with uncertain DRE are assessed by calculating the Expected Power Not Served (EPNS), and Expected Power Curtailment (EPC).

Assuming prediction errors of the DRE active power output follow a multivariate normal distribution, i.e., $P^t_{\text{DRE}} = \left(P^t_{\text{DRE}(1)}, \ldots, P^t_{\text{DRE}(m)}, \ldots, P^t_{\text{DRE}(M)}\right)^T \sim N\left(\mu^t_{\text{DRE}}, B^t\right)$, where M denotes the number of DREs. If α is a given real number and the confidence level of DRE is set to 95%, then $\alpha = 0.05$, $Z_{1-\alpha/2} = 1.96$, where Z represents the quantile of normal distribution, and the confidence interval for the DRE output can be formulated as follows:

$$\mu^t_{\text{DRE}} - Z_{1-\alpha/2}B^t \leq P^t_{\text{DRE}} \leq \mu^t_{\text{DRE}} + Z_{1-\alpha/2}B^t \tag{4.16}$$

where P^t_{DRE} denotes DRE output vector at time t, μ^t_{DRE} indicates the forecasted value vector of DRE, i.e., $\mu^t_{\text{DRE}} = \left(\mu^t_{\text{DRE}(1)}, \ldots, \mu^t_{\text{DRE}(m)}, \ldots, \mu^t_{\text{DRE}(M)}\right)^T$, and B^t refers to the covariance matrix of DRE at time t, i.e., $B^t = \text{Cov}\left(P^t_{\text{DRE}}\right)_{M\times M}$.

The joint probability distribution is used to derive the EPNS and EPC to address the forecasting error of DRE. Reserve power is then calculated based on the EPNS. If the DRE power output falls the confidence interval, the cost of the required reserve power can be determined by the first term of Eq. (4.8). Conversely, if the DRE's output exceeds the confidence interval, the surplus power will be sold to the upper grid with the on-grid electricity price.

The joint probability density function $\phi_J\left(P^t_{\text{DRE}}\right)$ and the joint cumulative distribution function $\Phi_J\left(P^t_{\text{DRE}}\right)$ of DRE output $P^t_{\text{DRE}} = \left(P^t_{\text{DRE}(1)}, \ldots, P^t_{\text{DRE}(m)}, \ldots, P^t_{\text{DRE}(M)}\right)^{\text{T}}$ are:

$$\phi_J\left(P^t_{\text{DRE}}\right) = \frac{1}{\sqrt{(2\pi)^M \left|B^t\right|}} e^{-\frac{1}{2}\left(P^t_{\text{DRE}} - \mu^t_{\text{DRE}}\right)^T \left(B^t\right)^{-1}\left(P^t_{\text{DRE}} - \mu^t_{\text{DRE}}\right)} \tag{4.17}$$

$$\Phi_J\left(P^t_{\text{DRE}}\right) = \int\limits_{-\infty}^{\mu^t_{\text{DRE}(1)} + Z_{1-\alpha/2}B^t} \cdots \int\limits_{-\infty}^{\mu^t_{\text{DRE}(M)} + Z_{1-\alpha/2}B^t} \phi_J\left(P^t_{\text{DRE}}\right) dP^t_{\text{DRE}(1)} \cdots dP^t_{\text{DRE}(M)} \tag{4.18}$$

Then, the EPNS and EPC of mth DRE can be expressed as:

$$P^t_{\text{EPNS}(m)} = \int\limits_0^{P^t_{\text{DRE}(m)}} \left(P^t_{\text{DRE}(m)} - x\right)\phi_{\text{J}}(x)\text{d}x \tag{4.19}$$

$$P^t_{\text{EPC}(m)} = \int\limits_{P^t_{\text{DRE}(m)}}^{\mu^t_{\text{DRE}(m)} + Z_{1-\alpha/2}B^t} \left(x - P^t_{\text{DRE}(m)}\right)\phi_{\text{J}}(x)\text{d}x \tag{4.20}$$

Note that $P^t_{\text{EPNS}(m)}$ represents that the active power output of mth DRE is less than the forecasted value, while $P^t_{\text{EPC}(m)}$ denotes that the active power output of the mth DRE is greater than the forecasted value.

Let $(Q^t_{\text{DRE}})^2 = C^t_{\text{DRE}} - (P^t_{\text{DRE}})^2$, i.e.,$P^t_{\text{DRE}} = \sqrt{C^t_{\text{DRE}} - (Q^t_{\text{DRE}})^2}$, where $C^t_{\text{DRE}} = \left(C^t_{\text{DRE}(1)}, \ldots, C^t_{\text{DRE}(m)}, \ldots, C^t_{\text{DRE}(M)}\right)^T = \min\{S^t_{\text{DRE}}, P^t_{\text{DRE}}/PF_{\text{DRE}}\}$. Then the Jacobi matrix of Q^t_{DRE} is:

$$J(Q^t_{\text{DRE}}) = \begin{vmatrix} \partial P^t_{\text{DRE}(1)}/\partial Q^t_{\text{DRE}(1)} & \partial P^t_{\text{DRE}(1)}/\partial Q^t_{\text{DRE}(2)} & \cdots & \partial P^t_{\text{DRE}(1)}/\partial Q^t_{\text{DRE}(M)} \\ \partial P^t_{\text{DRE}(2)}/\partial Q^t_{\text{DRE}(1)} & \partial P^t_{\text{DRE}(2)}/\partial Q^t_{\text{DRE}(2)} & \cdots & \partial P^t_{\text{DRE}(2)}/\partial Q^t_{\text{DRE}(M)} \\ \vdots & \vdots & \cdots & \vdots \\ \partial P^t_{\text{DRE}(M)}/\partial Q^t_{\text{DRE}(1)} & \partial P^t_{\text{DRE}(M)}/\partial Q^t_{\text{DRE}(2)} & \cdots & \partial P^t_{\text{DRE}(M)}/\partial Q^t_{\text{DRE}(M)} \end{vmatrix}$$
$$= \frac{Q^t_{\text{DRE}(1)} \cdots Q^t_{\text{DRE}(m)} \cdots Q^t_{\text{DRE}(M)}}{\sqrt{C^t_{\text{DRE}(1)} - (Q^t_{\text{DRE}(1)})^2} \cdots \sqrt{C^t_{\text{DRE}(m)} - (Q^t_{\text{DRE}(m)})^2} \cdots \sqrt{C^t_{\text{DRE}(M)} - (Q^t_{\text{DRE}(M)})^2}} \tag{4.21}$$

where $Q^t_{\text{DRE}} = \left(Q^t_{\text{DRE}(1)}, \ldots, Q^t_{\text{DRE}(m)}, \ldots, Q^t_{\text{DRE}(M)}\right)^T$ indicates DRE reactive power vector.

In consideration of the forecasting error in DRE active power output, the joint probability density function for DRE reactive power output Q^t_{DRE} can be written as:

$$l(Q^t_{\text{DRE}}) = \phi_J(P^t_{\text{DRE}})|J(Q^t_{\text{DRE}})| \tag{4.22}$$

Similarly, the EPNS of the mth DRE reactive power output is:

$$Q^t_{\text{EPNS}(m)} = \int_0^{Q^t_{\text{DRE}(m)}} (Q^t_{\text{DRE}(m)} - x)1(x)\mathrm{d}x \tag{4.23}$$

Note that $Q^t_{\text{EPNS}(m)}$ denotes that the reactive power output of mth DRE is less than the forecasted value.

Therefore, the reserve power required from micro-turbine m to address the DRE uncertainty can be represented as $\lambda^t_m E^t_{\text{g}}$. This corresponds to the first term of Eq. (4.8), where λ^t_m denotes the participation factor of reserve from micro-turbine m at time t, and the total reserve power E^t_{g} required is:

$$E^t_{\text{g}} = \sum_{m=1}^{M} \sqrt{(P^t_{\text{EPNS}(m)})^2 + (Q^t_{\text{EPNS}(m)})^2} \tag{4.24}$$

4.4 Many-Objective Reduction

When a model encompasses numerous objective functions, it can lead to a number of Pareto solutions, complicating the optimization and decision-making process. Consequently, it is crucial to analyze the interconnections among different objectives and reduce their number. The correlation between objectives x and y can be described by correlation coefficient ρ_{xy}:

$$\rho_{xy} = \frac{\mathrm{cov}\left(R_{x(\eta)}, R_{y(\eta)}\right)}{\sigma_{x(\eta)}\sigma_{y(\eta)}} \tag{4.25}$$

where $\sigma_{x(\eta)}$ and $\sigma_{y(\eta)}$ denote deviations of objectives x and y in the solution set η, respectively, $R_{x(\eta)}$ and $R_{y(\eta)}$ represent superiority and inferiority ranking values of objectives x and y in the solution set η, respectively. Therefore, in this chapter, the correlation coefficient matrix $R^{\mathrm{N}} = \left\{\rho_{xy}\right\}(1 \le x \le H, 1 \le y \le H)$ is employed to assess the relationship between H objectives. A positive " + " or negative "-" sign in ρ_{xy} indicates whether objectives x and y are in harmony or conflict, respectively, while the absolute value of ρ_{xy} reflects the strength of their correlation. An absolute value close to zero suggests that the objectives are relatively independent, whereas a value of + 1 or − 1 denotes a strict monotonic relationship between the objectives. This non-parametric measurement method does not require objectives to be comparable or transformed into a common unit.

To solve the proposed high-dimensional many-objective optimization, this chapter introduces the multiple preys-based evolutionary predator and prey strategy (MPEPPS). In the MPEPPS framework, the population is derived into experienced predators, strategic predators, preys, and safe positions. The detailed steps of MPEPPS for optimization are as follows:

Step 1: Optimize the initial set of objectives by the MPEPPS and then establish an initial Pareto solution. This involves the correlation coefficient matrix calculation and K-means clustering to generate a new composite set that integrates harmonious objectives. This new set, which consists of conflicting objectives, reduces dimensionality and enhances the optimization process without altering the shape of the original Pareto front.

Then, the optimal clusters are determined by the Calinski-Harabasz index. Calinski-Harabasz index assesses clustering effectiveness based on the distances between sample points and cluster centers within the dataset [30].

$$s = \frac{tr(B_k)(m_{\mathrm{k}} - k)}{tr(W_k)(k_k - 1)} \tag{4.26}$$

where B_k denotes distance between classes, W_k indicates distance within classes, m_k refers to the sample size, and k_k signifies the number of clusters.

The equations for calculating the distance between classes and the distance within classes are:

$$\begin{cases} B_k = \sum_{q=1}^{k} n_q (c_q - c_E)(c_q - c_E)^T \\ W_k = \sum_{q=1}^{k} \sum_{x_q \in C_q} (x_q - c_q)(x_q - c_q)^T \end{cases} \tag{4.27}$$

where n_q denotes the number of data points of class q, c_q indicates a particle of class q, c_E signifies the data center point, c_q refers to the set of data in class q, and x_q is the data in c_q.

Step 2: In this step, the variation coefficient is employed to identify objective values that exhibit minimal changes in optimization. This approach helps to reduce computational burden and prevent the inefficient use of resources. The variation coefficient is given by Eq. (4.28).

$$C_{cv} = \frac{\sigma_{cv}}{|\mu_{cv}|} \times 100\% \tag{4.28}$$

where C_{cv} denotes variation coefficient, σ_{cv} indicates the standard deviation, and μ_{cv} refers to the mean value of correlation coefficient.

Firstly, objectives with a variation coefficient less than 0.1%, while those with higher coefficients are retained as targets. The MPEPPS algorithm is then utilized to optimize this revised objective set, producing a new Pareto front. Following this, a new correlation coefficient matrix is calculated. The objective that exhibits the greatest conflict with other objectives is identified and removed from the set. The Pareto front of the modified objective set is compared with that of the set including the removed objective to assess the impact of the exclusion. This process is iterated until no further objectives can be eliminated.

Step 3: Calculate the correlation coefficient matrix of the remaining objectives. Then, cluster these objectives into two sets based on their conflicts. In view of the reduction, a many-objective optimization is applied to obtain the final Pareto solutions.

This chapter introduces an iterative bi-level optimization for stochastic DNR, which integrates multi-step reconstruction and many-objective reduction techniques. At the upper level, the initial iteration incorporates predicted values of DRE and micro-turbines. The DNR is performed using the MSSE to minimize power loss, ultimately determining the topology that results in the least power loss.

At the lower level, a many-objective reduction optimization is developed in consideration of the topology derived from the upper level. This model evaluates the risk caused by the uncertain output of DRE. The optimized active and reactive power outputs of the DRE and micro-turbines are then used as inputs for the upper level in the subsequent iteration. The flowchart of the proposed iterative bi-level scheduling for DNR is illustrated in Fig. 4.1.

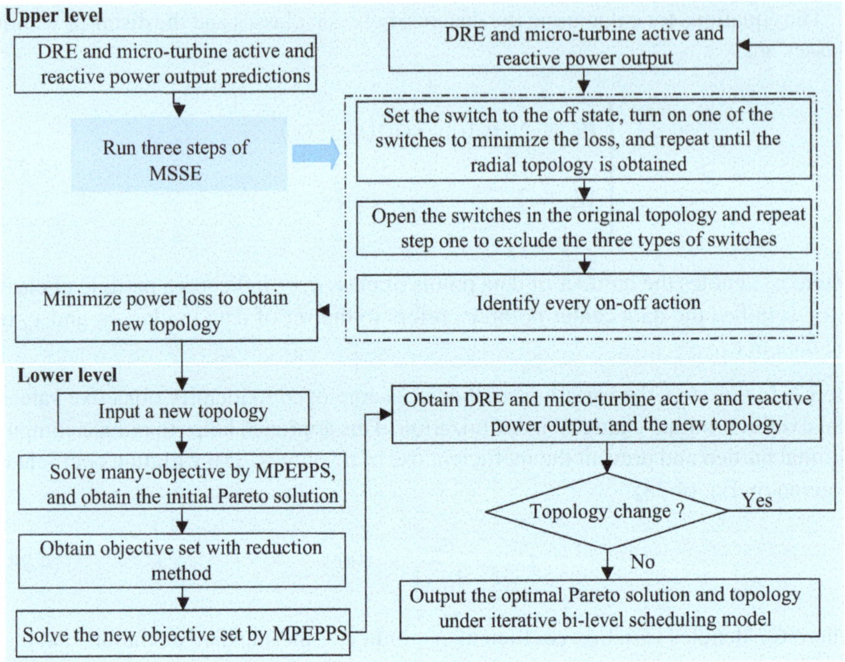

Fig. 4.1 Flowchart of iterative bi-level scheduling for stochastic distribution network Reprinted from Ref. [35]. Copyright 2024, with permission Energy

4.5 Case Study

To evaluate the proposed bi-level optimization, numerical studies are conducted on IEEE 33-node, 84-node, 119-node, and 136-node systems, using PV technology as a case study for DRE performance. For the IEEE 33-node system, nodes 8, 22, and 23 are equipped with DREs and the capacities are respectively 350 kW, 350 kW, and 150 kW, along with micro-turbines rated at 500 kW, 500 kW, and 300 kW. For the IEEE 84-node, IEEE 119-node, and IEEE 136-node systems, Voltage limits are set at 1.0 p.u. and 0.9 p.u., in accordance with the recommendations in [31]. The DRE's capacity is configured to 1.1 times the maximum active power, and the power factor is established at 0.9. The on-grid electricity price is 0.394 CNY/kWh, and the cost coefficients for the micro-turbines are assigned values of 0, with coefficients set at 1.2, 1.1, and 1.3, respectively. Due to the specific nature of PV, the analysis is limited to a 12-h period from 7:00 to 18:00. The generations and population size of MPEPPS are respectively 1000 and 200. To reduce statistical errors, each test is repeated 30 times independently.

Table 4.1 Power loss of each test system under different DNR methods

DNR method (kW)	IEEE 33-node	IEEE 84-node	IEEE 119-node	IEEE 136-node
MSSE	125.920	406.682	570.141	244.329
NTH	126.985	542.245	644.983	294.956
SOE	126.354	523. 721	640.322	287.288

4.5.1 Simulation Analysis of DNR Based on MSSE

In bi-level optimization, the optimization results from the lower-level optimization of DREs and micro-turbines are fed into the upper level. After several iterations, the optimal network topology is determined. The upper level utilizes a DNR method based on the MSSE approach to minimize power loss and identify the optimal topology. Then, MSSE method is implemented in three phases, with each phase yielding the minimum power loss and the corresponding topology. This subsection presents a comparison and analysis of power loss across four test systems based on optimization results from 8:00 AM.

To evaluate the MSSE, a thorough comparison is conducted with the SOE [10] and NTH [11]. The optimized power losses for different test system under thses three DNR are detailed in Table 4.1. In particular, for the IEEE 33-node system, the MSSE achieves power losses that are 1.065 kW and 0.434 kW lower than those obtained using NTH and SOE, respectively. MSSE demonstrates a marked reduction in power loss in larger node systems. Specifically, for the IEEE 84-node, IEEE 119-node, and IEEE 136-node systems, the MSSE results in power loss reductions of 25.00%, 11.60%, and 17.16% compared to the NTH, and 22.35%, 10.96%, and 14.95% compared to the SOE, respectively.

The MSSE1 achieves the lowest power loss on IEEE 33-node system, as shown in the second row of Table 4.2. This suggests that, for smaller node systems, the optimal topology can be determined in a single step. In the third, fourth, and fifth columns of the table, it is evident that the MSSE method consistently achieves lower power loss per step for the IEEE 84-node, IEEE 119-node, and IEEE 136-node systems. Specifically, the obtained power loss for MSSE3 in these three systems is reduced by 3.021 kW, 5.142 kW, and 2.126 kW compared to MSSE1, respectively. These findings verify the effectiveness of MSSE-based DNR for power loss reduction.

4.5.2 Simulation Analysis of Many-Objective Reduction

This subsection uses the IEEE 33-node system as a case study to investigate the objective reduction method. Initially, the MPEPPS is employed to determine the initial Pareto solutions for the six-objective functions, and a correlation coefficient

Table 4.2 Power loss of each system in different scenarios

System	MSSE0 (kW)	MSSE1 (kW)	MSSE2 (kW)	MSSE3 (kW)
IEEE 33-node	131.710	125.920	125.920	125.920
IEEE 84-node	628.175	409.703	408.752	406.682
IEEE 119-node	734.354	575.283	574.660	570.141
IEEE 136-node	364.154	246.455	245.665	244.329

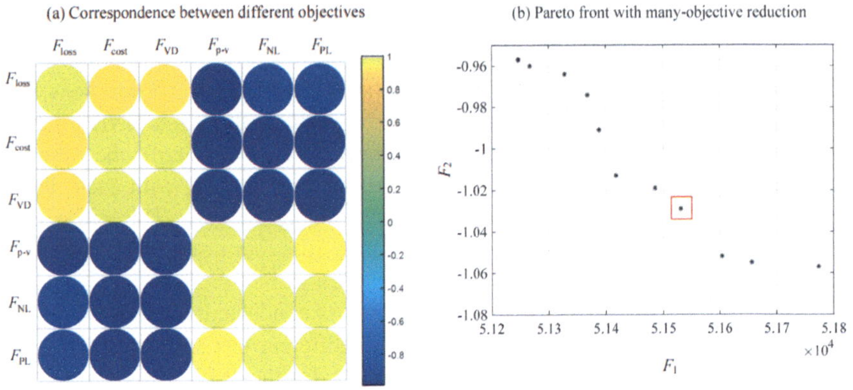

Fig. 4.2 The initial and optimal typologies of IEEE 33-node system Reprinted from Ref. [35]. Copy-right 2024, with permission energy

matrix is generated. Figure 4.2a illustrates the relationships among the objectives, revealing that these objectives are relatively distinct.

Specifically, the power loss, operation cost, and voltage deviation exhibit high positive correlations, with correlation coefficients exceeding 0.9. In contrast, these three objectives show a strong negative correlation with the net load factor, net load peak-valley difference, and power load rate, as evidenced by correlation coefficients below $-$ 0.9. Subsequently, the six objectives are categorized into two conflicting groups through K-clustering:

$$\min(F_1, F_2) \tag{4.29}$$

where $F_1 = F_{\text{cost}} + F_{\text{loss}} + F_{\text{VD}}$, $F_2 = F_{\text{p-v}} - F_{\text{NL}} - F_{\text{PL}}$. The coefficient variations of F_1 and F_2 are greater than 0.1%. Figure 4.2b depicts the Pareto front resulting from the many-objective reduction.

To select the optimal trade-off from Pareto solutions, the Technique for Order Preference by Similarity to Ideal Solution (TOPSIS) is used [32]. TOPSIS operates on the principle that the best alternatives are those closest to the positive ideal solution and farthest from the negative ideal solution [33]. Its applicability to our study is due to its straightforward calculations and its capacity to incorporate relative weights according to the importance of criteria [34]. A compromise solution is then selected, as shown in Fig. 4.2b, which is highlighted within the red box.

4.5.3 Simulation Analysis of Iterative Bi-level Scheduling

The lower level model is optimized by the proposed many-objective reduction, resulting in a simplified optimization with only two optimization objectives. The optimal active and reactive power outputs for DREs and micro-turbines were then used as inputs for the upper level model to refine the topology.

Simulation analysis is first conducted on the IEEE 33-node system. The original system topology, depicted in Fig. 4.3a, includes tie switches at 7–8, 9–10, 14–15, 25–29, and 32–33. After several iterations of the bi-level optimization, the optimal topology was achieved, as shown in Fig. 4.3b, with tie switches at 7–8, 9–10, 14–15, 28–29, and 18–33. Similarly, the IEEE 84-node, IEEE 119-node, and IEEE 136-node systems also reached their optimal topologies following the double-layer iteration process, as illustrated in Fig. 4.4. Additionally, Table 4.3 presents the optimal active and reactive power outputs for DREs and micro gas turbines in the IEEE 33-node system based on the topology shown in Fig. 4.3b. Notably, the double-layer iteration for the IEEE 33-node system took 222.64 s in total, with 90.46 s allocated to the upper layer and 132.18 s to the lower layer.

To assess the bi-level stochastic optimization, two scenarios are examined: Scenario 1 (S1), which does not incorporate DNR, and Scenario 2 (S2), which employs the bi-level stochastic optimization. Table 4.3 presents the optimal results

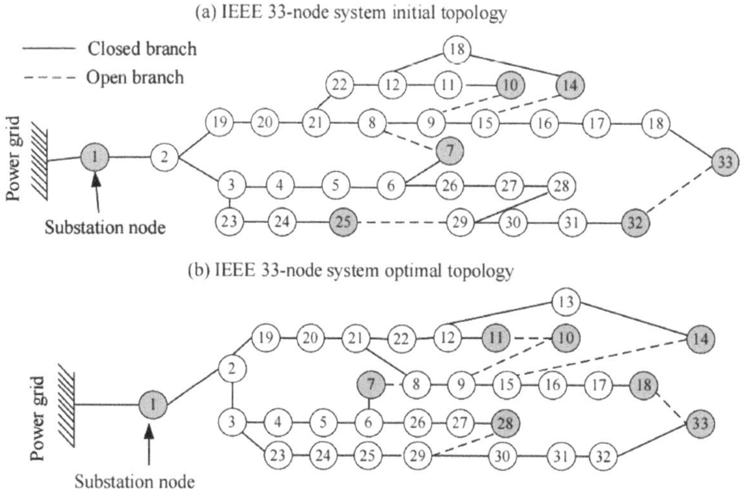

Fig. 4.3 The initial and optimal typologies of IEEE 33-node system Reprinted from Ref. [35]. Copy-right 2024, with permission Energy

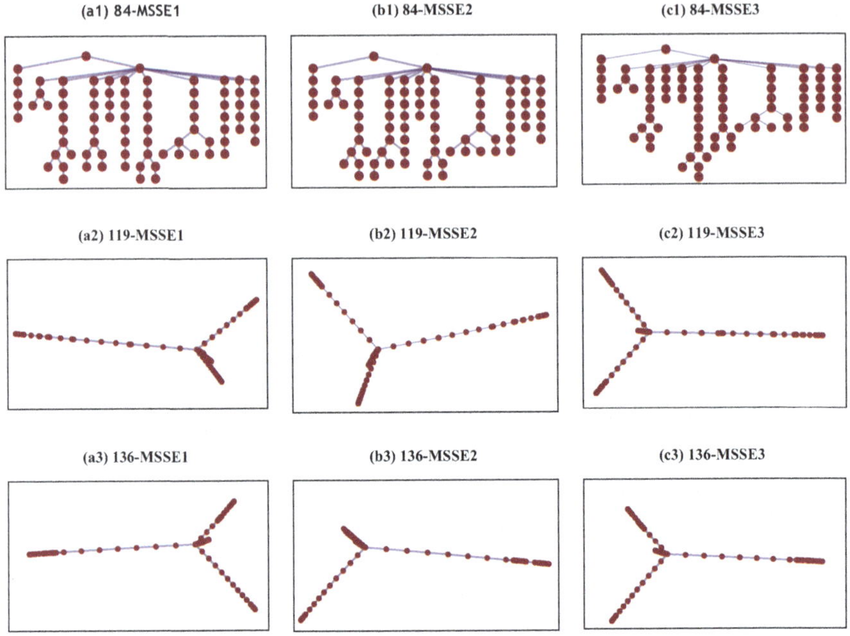

Fig. 4.4 The optimal typologies of IEEE 84-node, IEEE 119-node, and IEEE 136-node systems

obtained under these two scenarios for the IEEE 33-node system. The analysis reveals that Scenario 2 achieves an operational cost reduction of 201.74 CNY compared to Scenario 1. Additionally, voltage deviation and power loss are significantly improved in Scenario 2, with reductions of 13.05% and 10.13%, respectively. The net load peak-valley difference in Scenario 2 is reduced by 0.021 compared to Scenario 1, while the net load factor and power load rate increase by 0.003 and 0.085, respectively. Overall, the results for Scenario 2 demonstrate superior performance relative to Scenario 1, highlighting that the proposed iterative bi-level stochastic scheduling effectively lowers operational costs and power loss, enhances voltage quality, and improves net load adaptation and equipment utilization.

Figure 4.5 illustrates the power loss for the IEEE 33-node system under both S1 and S2. The figure shows that power loss in Scenario 2 is consistently lower than in Scenario 1 at all times. Peak energy consumption occurs between 9:00–10:00 and 15:00–18:00, periods characterized by high load demand and correspondingly high power loss. Conversely, the valley period, occurring between 12:00 and 14:00, shows low load demand and a minimal power loss of 80.047 kW at 12:00. These findings indicate that the proposed optimization effectively reduces power loss and enhances the operational efficiency of the distribution network (Table 4.4).

Figure 4.6 depicts the voltage distributions for each node of Scenarios 1 (S1) and 2 (S2) at various times. Both scenarios maintain node voltages within the prescribed limits, with no violations. Notably, Fig. 4.6b shows less voltage fluctuation compared

Table 4.3 Optimal active and reactive power of DRE and micro-turbine in IEEE-33 node

Time (h)	$P_{DRE,1}$(kW)	$P_{DRE,2}$ (kW)	$P_{DRE,3}$ (kW)	$Q_{DRE,1}$ (kW)	$Q_{DRE,2}$ (kW)	$Q_{DRE,3}$ (kW)
7	112.143	115.985	65.420	− 58.237	22.722	11.433
8	170.076	160.640	107.816	31.101	− 25.236	28.083
9	142.170	224.075	106.522	− 144.967	88.536	− 25.839
10	340.141	199.504	119.931	− 16.955	50.221	81.498
11	311.664	313.864	118.288	159.266	− 61.346	92.238
12	240.462	311.989	80.599	81.364	103.206	− 102.639
13	332.436	332.640	95.768	109.481	7.132	− 115.449
14	222.134	329.827	147.191	− 49.585	− 117.106	28.893
15	272.541	249.818	122.570	− 58.109	− 183.767	86.468
16	280.659	228.984	118.643	− 0.119	− 115.337	− 14.214
17	129.202	199.058	120.192	115.771	− 241.864	50.940
18	122.418	143.982	51.674	4.916	64.827	− 55.351
Time (h)	$P_{g,1}$ (kW)	$P_{g,2}$ (kW)	$P_{g,3}$ (kW)	$Q_{g,1}$ kW)	$Q_{g,2}$ (kW)	$Q_{g,3}$ (kW)
7	4.171	214.965	227.396	95.146	251.820	219.196
8	56.580	241.584	107.755	194.768	198.526	30.578
9	444.263	161.843	49.424	428.782	337.112	45.031
10	188.704	375.642	367.938	343.298	327.058	155.287
11	344.955	38.500	481.209	227.161	250.977	86.131
12	213.718	135.017	131.908	215.900	82.012	66.549
13	82.546	116.759	12.327	44.952	134.829	67.099
14	109.139	378.257	176.932	487.981	300.658	28.802
15	216.965	279.711	104.467	112.784	180.465	142.783
16	228.611	60.886	93.515	154.642	217.356	0.523
17	70.408	232.915	60.199	286.333	176.418	98.490
18	37.791	266.372	127.804	160.106	23.942	73.632

Table 4.4 Optimal objective values for different scenarios in IEEE-33 node system

Scenario	Total cost (CNY)	Power loss(kW)	Voltage deviation	Net load peak-valley difference	Net load coefficient	Power load ratio
S1	49,894.80	2030.298	15.569	0.493	0.709	0.703
S2	49,693.06	1824.556	13.537	0.472	0.712	0.788

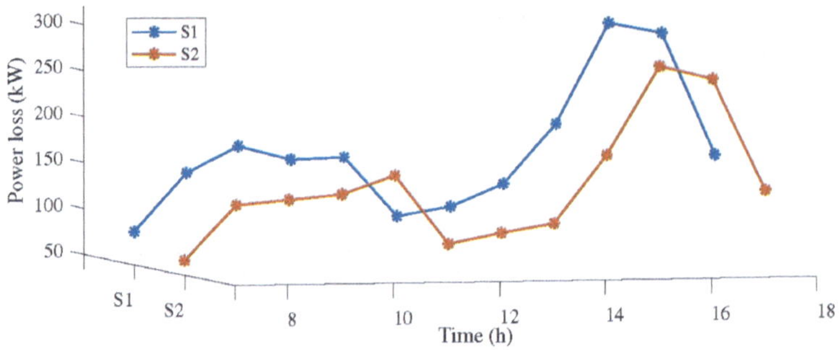

Fig. 4.5 The power loss of IEEE 33-node system based on S1 and S2 Reprinted from Ref. [35]. Copy-right 2024, with permission Energy

to Fig. 4.6a. The minimum voltage in Scenario 1 is recorded at node 33 at 16:00, with a value of 0.909 p.u., whereas in Scenario 2, the minimum voltage is recorded at node 32 at the same time, which is 0.908 p.u., representing an improvement of 0.01 p.u. This suggests that the proposed bi-level stochastic optimization effectively enhances voltage quality.

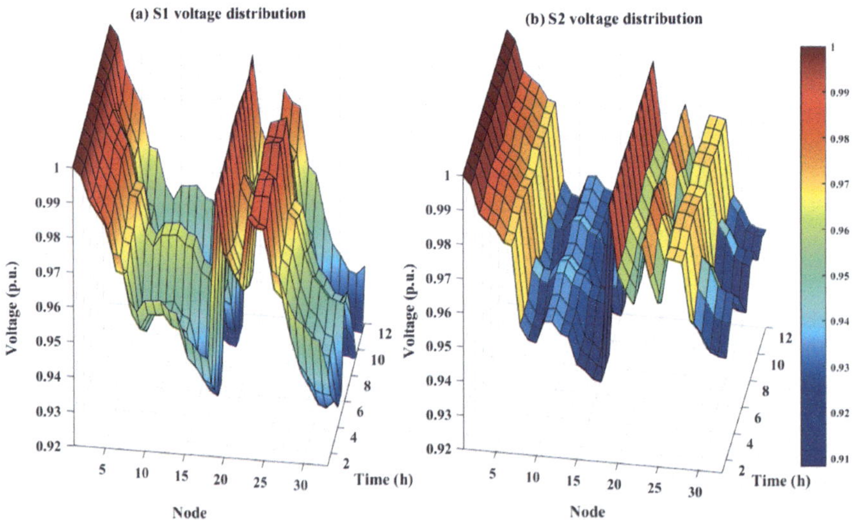

Fig. 4.6 The voltage distribution of each node at different times of S1 and S2

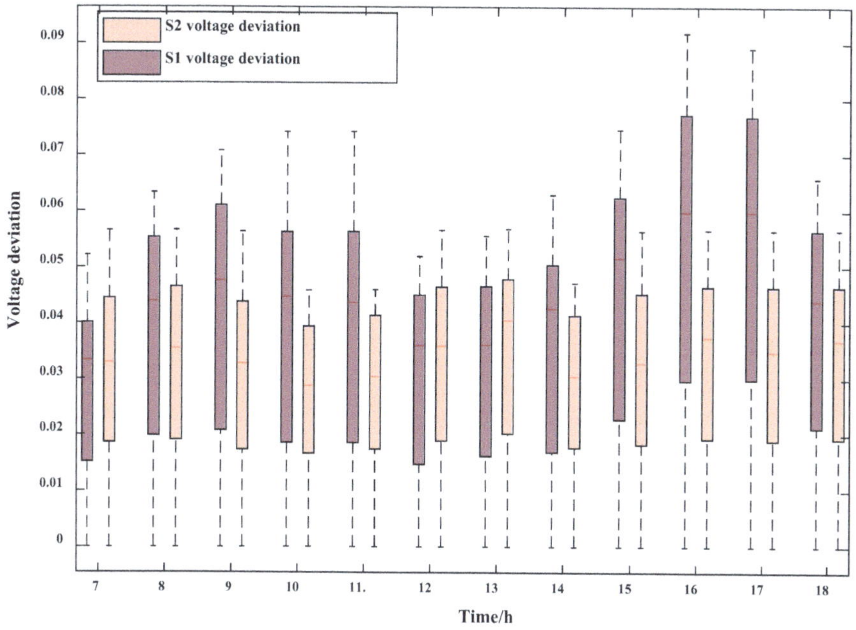

Fig. 4.7 The voltage deviation of S1 and S2

In addition, Fig. 4.7 presents a box plot of voltage deviations for both scenarios. The plot indicates that Scenario 2 generally exhibits smaller voltage deviations than Scenario 1. Additionally, the median value of voltage deviation at each node in Scenario 2 is consistently lower than in Scenario 1. The most pronounced difference in voltage deviation between the two scenarios occurs at 16:00. This underscores the superior voltage quality achieved with the bi-level optimization, denoted as Scenario 2. Therefore, the proposed optimization can both reduce power loss and voltage deviation.

4.6 Conclusion

This chapter introduces an iterative bi-level stochastic scheduling for DNR, integrating multi-step reconfiguration and many-objective reduction. The simulation results conducted on several test systems indicate the following:

(1) The MSSE reconfiguration approach significantly outperforms the SOM and SOE methods in reducing power loss. The reduction rate with MSSE ranges from 11.60 to 25.00%, whereas with SOM and SOE, it ranges from 10.96 to 22.35%.

(2) The objective reduction effectively reduces computational complexity by condensing six objectives into two. In addition, this method maintains node voltages within the range of 0.908 to 1.0, with the minimum acceptable voltage set at 0.90 under MSSE reconfiguration.

(3) The proposed optimization was comprehensively studied to prove its effective in reducing operational costs, voltage deviations, and network losses, while also aiding in peak shaving and valley filling.

References

1. Mansouri SA, Ahmarinejad A, Nematbakhsh E, Javadi MS, Nezhad AE, Catala˜o JP (2022) A sustainable framework for multi-microgrids energy management in automated distribution network by considering smart homes and high penetration of renewable energy resources. Energy 245:123228–53
2. Li Y, Feng B, Wang B, Sun S (2022) Joint planning of distributed generations and energy storage in active distribution networks: a bi-level programming approach. Energy 245:123226–123249
3. Kennedy KM, Ruggles TH, Rinaldi K, Dowling JA, Duan L, Caldeira K et al (2022) The role of concentrated solar power with thermal energy storage in least-cost highly reliable electricity systems fully powered by variable renewable energy. Adv Appl Energy 6(100091)
4. Yang Z, Yang F, Min H, Tian H, Hu W, Liu J et al (2023) Energy management programming to reduce distribution network operating costs in the presence of electric vehicles and renewable energy sources. Energy 263:125695–125706
5. Gao H, Ma W, He S, Wang L, Liu J (2022) Time-segmented multi-level reconfiguration in distribution network: a novel cloud-edge collaboration framework. IEEE Trans Smart Grid 13(4):3319–3322
6. Azizivahed A, Narimani H, Fathi M, Naderi E, Safarpour HR, Narimani MR (2018) Multi-objective dynamic distribution feeder reconfiguration in automated distribution systems. Energy 147:896–914
7. Merlin A (1975) Search for a minimal-loss operating spanning tree configuration for an urban power distribution system. In: Proceedings of 5th PSCC, vol 1, pp 1–18
8. Singh S, Raju G, Rao G, Afsari M (2009) A heuristic method for feeder reconfiguration and service restoration in distribution networks. Int J Electr Power Energy Syst 31(7–8):309–314
9. Ghasemi S, Moshtagh J (2014) A novel codification and modified heuristic approaches for optimal recon- figuration of distribution networks considering losses cost and cost benefit from voltage profile improvement. Appl Soft Comput 25:360–368
10. Zhan J, Liu W, Chung C, Yang J (2020) Switch opening and exchange method for stochastic distribution network reconfiguration. IEEE Trans Smart Grid 11(4):2995–3007
11. Harsh P, Das D (2023) A simple and fast heuristic approach for the reconfiguration of radial distribution networks. IEEE Trans Power Syst
12. Jahani MTG, Nazarian P, Safari A, Haghifam M (2019) Multi-objective optimization model for optimal reconfiguration of distribution networks with demand response services. Sustain Cities Soc 47:101514–101525
13. Wang HJ, Pan JS, Nguyen TT, Weng S (2022) Distribution network reconfiguration with distributed generation based on parallel slime mould algorithm. Energy 244:123011–123030
14. Azizivahed A, Narimani H, Naderi E, Fathi M, Narimani MR (2017) A hybrid evolutionary algorithm for secure multi-objective distribution feeder reconfiguration. Energy 138:355–373
15. Zhang Y, Qian T, Tang W (2022) Buildings-to-distribution-network integration considering power transformer loading capability and distribution network reconfiguration. Energy 244:123104–123120

16. Lannoye E, Flynn D, O'Malley M (2012) Evaluation of power system flexibility. IEEE Trans Power Syst 27(2):922–931
17. Tian Y, Si L, Zhang X, Cheng R, He C, Tan KC et al (2021) Evolutionary large-scale multi-objective optimization: a survey. ACM Comput Surv (CSUR) 54(8):1–34
18. Qin Y, Zheng J, Wu Q (2021) Many-objective interactive optimization and decision making for distribution network expansion planning. Control Eng Pract 116:104917–104926
19. Wang H, Yao X (2016) Objective reduction based on nonlinear correlation information entropy. Soft Comput 20:2393–2407
20. de Freitas AR, Fleming PJ, Guimaraes FG (2015) Aggregation trees for visualization and dimension reduction in many-objective optimization. Inf Sci 298:288–314
21. Li H, Wang Z, Hong T, Piette MA (2021) Energy flexibility of residential buildings: a systematic review of characterization and quantification methods and applications. Adv Appl Energy 3(100054)
22. Kou Y, Zheng J, Li Z, Wu Q (2017) Many-objective optimization for coordinated operation of integrated electricity and gas network. J Modern Power Syst Clean Energy 5(3):350–363
23. Liu J, Chen J, Yan G, Chen W, Xu B (2023) Clustering and dynamic recognition based auto-reservoir neural network: a wait-and-see approach for short-term park power load forecasting. Iscience 26(107456)
24. Wang W, Huang Y, Yang M, Chen C, Zhang Y, Xu X (2022) Renewable energy sources planning considering approximate dynamic network reconfiguration and nonlinear correlations of uncertainties in distribution network. Int J Electr Power Energy Syst 139:107791–107801
25. Cui Z, Bai X, Li P, Li B, Cheng J, Su X et al (2020) Optimal strategies for distribution network reconfiguration considering uncertain wind power. CSEE J Power Energy Syst 6(3):662–671
26. Larimi SMM, Haghifam MR, Moradkhani A (2016) Risk-based reconfiguration of active electric distribution networks. IET Gener Transm Distrib 10(4):1006–1015
27. Jiao P, Chen J, Cai X, Wang L, Zhao Y, Zhang X et al (2021) Joint active and reactive for allocation of renewable energy and energy storage under uncertain coupling. Appl Energy 302:117582–117600
28. Li Y, Hao G, Liu Y, Yu Y, Ni Z, Zhao Y (2021) Many-objective distribution network reconfiguration via deep reinforcement learning assisted optimization algorithm. IEEE Trans Power Deliv 37(3):2230–2244
29. Zhong T, Zhang HT, Li Y, Liu L, Lu R (2020) Bayesian learning-based multi-objective distribution power network reconfiguration. IEEE Trans Smart Grid 12(2):1174–1184
30. Xing L, Sela L (2019) Unsteady pressure patterns discovery from high-frequency sensing in water distribution systems. Water Res 158:291–300
31. Wu H, Dong P, Liu M (2018) Distribution network reconfiguration for loss reduction and voltage stability with random fuzzy uncertainties of renewable energy generation and load. IEEE Trans Ind Inf 16(9):5655–5666
32. Li Y, Wang J, Zhao D, Li G, Chen C (2018) A two-stage approach for combined heat and power economic emission dispatch: Combining multi-objective optimization with integrated decision making. Energy 162:237–254
33. Wu P, He Y, Li Y, He J, Liu X, Wang Y (2022) Multi-objective optimisation of machining process parameters using deep learning-based data-driven genetic algorithm and topsis. J Manuf Syst 64:40–52
34. Deb K, Pratap A, Agarwal S, Meyarivan T (2002) A fast and elitist multiobjective genetic algorithm: NSGA- II. IEEE Trans Evol Comput 6(2):182–197
35. Li JY, Chen JJ, Wang YX et al (2024) Combining multi-step reconfiguration with many-objective reduction as iterative bi-level scheduling for stochastic distribution network. Energy 290:130198

Chapter 5
Credibility Theory Based Fuzzy Chance-Constrained AC OPF for Renewable Power System

5.1 Introduction

One of the essential tools in power system analysis is Optimal Power Flow (OPF) [1, 2]. Several critical tasks of power system operation, such as unit commitment, market clearing, and security assessment, rely on the solution of the OPF [3, 4]. In recent years, due to the intermittent, fluctuating, and uncertain nature of Renewable Energy Sources (RESs), the challenge of forecast uncertainty has become a significant factor in OPF [5, 6]. Consequently, in cases where OPF decisions involve uncertain RESs, significant forecasting errors may lead to frequent violations of system. Therefore, addressing the uncertainty in RESs generation forecasts within the framework of OPF has become an imperative issue.

To tackle this challenge, extensive research efforts have been undertaken. Robust optimization emerges as a viable approach to handle uncertain RESs in OPF. For example, Gazijahani et al. [7] propose an adaptive robust optimization model, employing an alternating direction method to pinpoint worst-case RESs scenarios. However, it's important to note that robust optimization does not tolerate constraint violations under any uncertainty scenario, often resulting in overly conservative solutions [8, 9]. Another commonly adopted method for uncertain optimization problems is stochastic optimization, offering operators flexibility to balance cost and security considerations [10]. In their work, Dvorkin et al. [11] introduce a preventive security-constrained stochastic OPF model to dispatch controllable generators at minimum cost. Nonetheless, it should be acknowledged that stochastic optimization necessitates repeated sampling, which can impact solution efficiency.

In the past few years, chance-constrained optimization has garnered great attention in the realm of RESs integrated OPF. This approach necessitates constraints on violations at a predetermined confidence level, leading to solutions that are comparatively less conservative than those derived from robust optimization methods [12]. Notably, researchers have developed several models of chance-constrained DC OPF (CC-DCOPF) based on the linear approximation [13–15].

J. Chen and Y. Li, *Renewable Power System Optimization*, Smart Energy Systems,
https://doi.org/10.1007/978-981-97-8132-4_5

In the work by Peña et al. [16], an affine-based chance-constrained OPF was developed, ensuring a high probability that no line will exceed its rating. Roald et al. [17] introduce a method within the chance-constrained OPF framework to extend the use of corrective control of power flow devices, such as phase-shifting transformers, in response to uncertainty, thereby ensuring that constraints are met with a desired probability. Varawala et al. [18] propose a CC-DCOPF model that addresses forecast uncertainty in RESs generation by incorporating energy storage systems, leading to reduced operating costs and enhanced system operational efficiency.

However, it is essential to note that the CC-DCOPF formulation based on DC approximation may not be directly applicable to distribution system operations, as it assumes voltage magnitudes at rated values. A more suitable alternative is the chance-constrained AC OPF (CC-ACOPF) formulation, as presented in [19]. Venzke et al. [20] further develop a CC-ACOPF formulation by leveraging Semi-Definite Programming (SDP) relaxation for the AC power flow, introducing a sample-based approach to determine appropriate corrective control policies for voltage regulation.

In the context of distribution grids, Mieth et al. [21, 22] employ a derived CC-ACOPF with LinDistFlow linearized equations [23] to ensure compliance with nodal voltage limits. This approach underscores the significance of tailored formulations to address the specific operational constraints and objectives encountered in distribution systems.

However, the majority of existing works, such as those by Peña et al. [16], Roald et al. [17], Varawala et al. [18], Chen et al. [19], Venzke et al. [20], and Mieth et al. [21], make the assumption that RESs outputs follow Gaussian distributions. This assumption is made due to the ease with which the chance constraint can be reformulated into a second-order cone form under Gaussian distribution.

It has been highlighted in the works of Dvorkin et al. [11] and Haupt et al. [24] that empirical measurements of solar and wind power generation often exhibit asymmetry. This means that the Gaussian assumption, which is a symmetry assumption, might lead to a misestimation of the impact of forecast errors, especially from nodes with significant asymmetries. Such misestimations could result in ineffective and inefficient operational decisions.

Moreover, the research by Liu et al. [25] indicates that in addition to randomness, uncertainty in RESs also encompasses fuzziness. Liu et al. establish a mathematical foundation for this fuzziness based on the principles of credibility theory. Despite this, existing CC-OPF predominantly focuses on the stochastic characteristics of RESs uncertainty. Only limited efforts have been made to explore the potential of leveraging fuzzy chance-constrained model to handle RESs uncertainty, particularly considering its fuzzy characteristics. This presents an avenue for further research and development in the field, aiming to enhance the robustness and effectiveness of decision-making processes in power system operations.

To address the aforementioned research gaps, this study leverages the probable range of RESs outputs to address its fuzzy characteristics. It establishes a Fuzzy Chance-Constrained AC Optimal Power Flow (FCC-ACOPF) model based on credibility theory, capable of handling RESs uncertainty with asymmetrical distribution. The contributions of this study are outlined as follows:

(1) This study introduces a novel FCC-ACOPF model framework aimed at mitigating uncertain power fluctuations within the distribution system. This enables Distribution System Operators (DSOs) to make informed decisions by balancing risks and costs through the selection of credibility levels.
(2) The proposed model specifically addresses nodes characterized by significant asymmetry in the distribution of prediction errors. By doing so, it provides power flow solutions that adhere to prescribed credibility levels. This approach helps to avoid the pitfalls of invalid and inefficient operational decisions that may arise from overly simplistic symmetry assumptions.
(3) This study derives deterministic linearization expressions for voltage, active power, and reactive power within the framework of fuzzy chance constraints. These formulas, grounded in credibility theory, facilitate the solution of the model using existing optimization solvers.

The subsequent sections of this chapter are organized as follows: Sect. 5.2 presents the model and formulation of OPF. Then, Sect. 5.3 discusses the fuzzy chance-constrained AC OPF based on credibility theory. Section 5.4 gives the solving strategy for the proposed FCC-ACOPF. After that, case studies and related analyses are provided in Sect. 5.5. Finally, Sect. 5.6 summarizes the findings and contributions of the study.

5.2 Problem Description of OPF

5.2.1 Overview of Credibility Theory

Credibility theory offers a robust fuzzy theory framework based on possibility and necessity measures. This theory is utilized to assess the credibility of fuzzy events, where an event with a credibility of one is guaranteed to occur, and an event with a credibility of zero is guaranteed not to occur [26].

The credibility measure is introduced through the minimum supremum of variables within a fuzzy event set, as outlined by [25, 27]. For $\forall A \in \mathcal{R}$, the credibility measure referring to fuzzy variable $\xi \in A$ is expressed as:

$$\mathrm{Cr}(\xi \in A) = \left[\sup_{x \in A} \mu(x) + 1 - \sup_{x \in A^c} \mu(x)\right]/2 \tag{5.1}$$

where $\text{Cr}(\cdot)$ denotes credibility measure symbol, $\mu(x)$ represents the membership function, and $\sup(\cdot)$ indicates a symbol representing the supremum. The credibility measure are with the following four axioms:

Axiom 1: For a non-empty set $\Theta \in \mathcal{R}$, $\text{Cr}\{\Theta\} = 1$.
Axiom 2: $\text{Cr}\{A_1\} \leq \text{Cr}\{A_2\}$, for $A_1 \subseteq A_2 \subseteq \Theta$.
Axiom 3: $\text{Cr}\{A\} + \text{Cr}\{A^c\} = 1$, for $\forall A \subseteq \Theta$.
Axiom 4: $\text{Cr}\{\cup_i A_i\} = \sup_i \text{Cr}\{A_i\}$, for any set, there is $\sup_i \text{Cr}\{A_i\} < 0.5$,

The introduction of credibility theory establishes a theoretical foundation for managing the uncertainty associated with RESs and for delivering reliable solutions in OPF problems involving RES integration.

5.2.2 Preliminaries

This chapter studies a radial distribution system with $n + 1$ nodes, which is shown in Fig. 5.1.$\mathcal{N}$ denotes the node-set of the distribution network, indexed by $i \in \mathcal{N} = \{0, 1, 2, \cdots, n\}$. Let $\mathcal{N}^+$ be a node-set composed of non-root nodes, indexed by $k \in \mathcal{N}^+ = \{1, 2, \cdots, n\}$. The root node, also known as the substation, serves as the connection between the transmission system and the distribution system. $\mathcal{A}_i$ is defined as the ancestor node-set of node i in the radial system, and $\mathcal{C}_i$ is set of children nodes of node i.

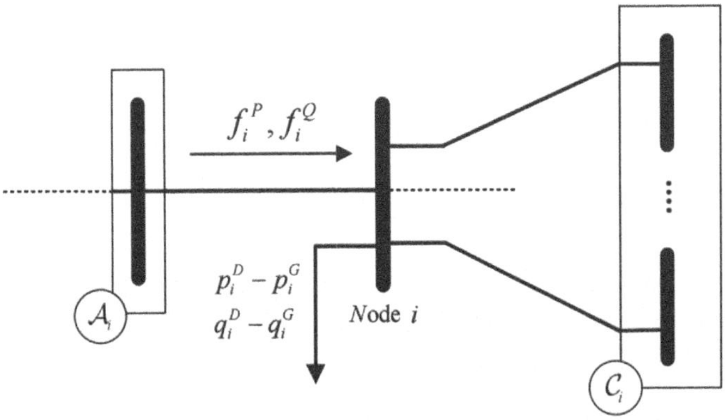

Fig. 5.1 Part of radial distribution system [22]

Let $\mathcal{G}$ be the node-set installed with controllable generator. For simplicity, we assume $\mathcal{G} = \mathcal{N}$. This assumption does not affect generality, as nodes without controllable generators can be accommodated by setting their output to zero. Consequently, the active power of the controllable generators is represented by the vector $\mathbf{p^G}$ indexed as p_i^G for $i \in \mathcal{G}$. Similarly, let vectors $\mathbf{d^P}$ indexed as d_i^P and $\mathbf{p^U}$ indexed as p_i^U. The reactive power vectors of controllable generator, node load and RESs are, respectively, indicated as $\mathbf{q^G}$,$\mathbf{d^Q}$,$\mathbf{q^U}$, and relationship of them are then expressed as:

$$\begin{aligned} p &= p^G - \left(d^P - p^U\right) \\ q &= q^G - \left(d^Q - q^U\right) \end{aligned} \tag{5.2}$$

The net load vector $\mathbf{p^D}$, indexed by p_i^D with $\mathbf{p^D} = \mathbf{d^P} - \mathbf{p^U}$. The voltage amplitude of node i is denoted by v_i, and this chapter uses the LinDistFlow for AC power flow, so that we set $u_i = v_i^2$. u_0 represents the base voltage of the substation on the secondary side, that is, $u_0 = 1.0$ p.u..

In Fig. 5.1, f_i^P,f_i^Q denote the active and reactive power flowing into node i, respectively. $\mathcal{L}$ represents lines set of the radial system, and $l = |\mathcal{L}|$ indicates the number of lines.

5.2.3 Uncertainty Model and Real-Time Balancing Regulation

The forecast error is an inherent issue due to the inherent unpredictability of RES generation. To address this, the forecast error of the net load is modeled by a fuzzy vector $\boldsymbol{\omega}$ indexed by ω_i for node i, and net load is represented as $\mathbf{p^D}(\omega) = \mathbf{p^D} + \boldsymbol{\omega}$. The forecast error referred to reactive power of node i can be expressed by $\omega_i^q = \omega_i \tan \varphi_i$ with $\cos \varphi_i = p_i^D / [\sqrt{(p_i^D)^2 + (q_i^D)^2}]$.

This chapter utilizes a trapezoidal fuzzy distribution to model the forecast error, as its ability to maintain a maximum membership within a certain range makes it suitable for representing uncertainty scenarios [28]. When the maximum degree of membership is sufficiently short, it approximates a triangular fuzzy number. Consequently, the membership function μ for the forecast error at node i is described as follows:

$$\mu(\omega_i) = \begin{cases} \frac{\omega_i - r_{i1}}{r_{i2} - r_{i1}}, & r_{i1} \leqslant \omega_i \leqslant r_{i2} \\ 1, & r_{i2} \leqslant \omega_i \leqslant r_{i3} \\ \frac{\omega_i - r_{i4}}{r_{i3} - r_{i4}}, & r_{i3} \leqslant \omega_i \leqslant r_{i4} \\ 0, & \text{otherwise} \end{cases} \tag{5.3}$$

where $r_{i1} < r_{i2} \leq r_{i3} < r_{i4}$ denote the fuzzy parameters used to describe the forecast error associated with the net load of node i. These parameters can generally be derived from studying historical data. Subsequently, the credibility distribution derived to address the forecast error, as expressed in Eq. (5.4), can be obtained using the credibility measure formula discussed in Eq. (5.1).

$$\mathrm{Cr}\{\xi \leq \omega_i\} = \begin{cases} 0, & \omega_i \leq r_{i1} \\ \frac{\omega_i - r_{i1}}{2(r_{i2} - r_{i1})}, & r_{i1} \leq \omega_i \leq r_{i2} \\ \frac{1}{2}, & r_{i2} \leq \omega_i \leq r_{i3} \\ \frac{\omega_i + r_{i4} - 2r_{i3}}{2(r_{i4} - r_{i3})}, & r_{i3} \leq \omega_i \leq r_{i4} \\ 1, & r_{i4} \leq \omega_i \end{cases} \tag{5.4}$$

The credibility distribution, as presented in Eq. (5.4), quantifies the credibility of the fuzzy variable ξ, where its value is less than or equal to a given threshold ω_i. Figure 5.2 illustrates the membership function and credibility distribution related to the forecast error of the net load.

In actual system operations, maintaining a well balance between power generation and demand amidst uncertain wind power is essential. To address this challenge, the linear affine strategy was employed, as discussed in Reference [29]. This strategy derives a proportional response from all participating generators to fluctuations in wind power. The mismatched power resulting from uncertain wind power was distributed among the generators according to their participation factors. Let $\boldsymbol{\alpha}$, indexed by α_i for $i \in \mathcal{G}$, represent the participation factor for each generator. The active and reactive power of each controllable generator is formulated as follows:

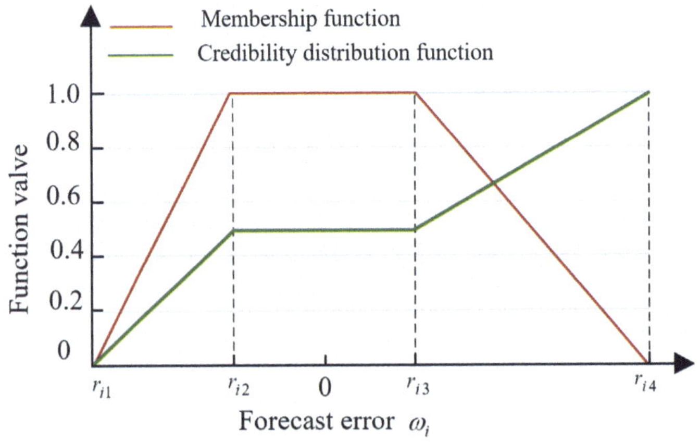

Fig. 5.2 Membership function and credibility distribution function

$$p_i^G(\omega) = p_i^G + \alpha_i \Omega^P$$
$$q_i^G(\omega) = q_i^G + \alpha_i \Omega^Q \tag{5.5}$$

where $\Omega^P = \sum_{i \in \mathcal{N}} \omega_i$ represents the total active power mismatch (Ω^Q by analogy). p_i^G,q_i^G denote the scheduled active and reactive power from the ith controllable generator, respectively. It is important to note that the participation factor is optimized concurrently with the scheduled generation.

Subsequently, the power flows and voltage amplitudes in the system respond to these adjustments according to the system's control mechanisms and physical relationships. Consequently, they will also vary with the forecast error and are represented as $v(\omega), f^P(\omega), f^Q(\omega)$, respectively.

5.3 FCC-ACOPF with LinDistFlow

The generation cost of generator i can be formulated as follows [30]:

$$c_i(p_i^G) = c_{2,i}(p_i^G)^2 + c_{1,i}p_i^G + c_{0,i} \tag{5.6}$$

where $c_{2,i}$, $c_{1,i}$ and $c_{0,i}$ are cost coefficients of generator i.

Given the forecasting error $\boldsymbol{\omega}$, and the credibility mean and variance of a fuzzy variable share properties similar to those of a random variable [31]. Thus, the fuzzy mean expression for the power generation cost can be derived as follows:

$$\begin{aligned}\mathbb{E}[c_i(p_i^G(\omega))] &= \mathbb{E}[c_{2,i}(p_i^G + \alpha_i \Omega^P)^2 + c_{1,i}(p_i^G + \alpha_i \Omega^P) + c_{0,i}] \\ &= c_i(p_i^G) + (2c_{2,i}p_i^G\alpha_i + c_{1,i}\alpha_i)\mathbb{E}(\Omega^P) + c_{2,i}\alpha_i^2\left[\mathbb{E}(\Omega^P)^2 + D(\Omega^P)\right]\end{aligned} \tag{5.7}$$

where $\mathbb{E}(\Omega^P)$ and $D(\Omega^P)$ denote mean and variance of the mismatched active power, respectively. Under the symmetry assumption, $\mathbb{E}(\Omega^P) = 0$. Consequently, the expected operational cost is simplified to:

$$\mathbb{E}\left[c_i\left(p_i^G(\omega)\right)\right] = c_i\left(p_i^G\right) + c_{2,i}\alpha_i^2 D(\Omega^P) \tag{5.8}$$

When accounting for the forecast error of RES power, the FCC-ACOPF can be expressed as follows:

$$
\begin{aligned}
&\min \quad \sum_{i\in\mathcal{G}} \mathbb{E}\left[c_i\left(p_i^G(\omega)\right)\right] && (5.9a)\\
&\text{s.t.} \quad f_i^P(\omega) + p_i^G(\omega) - \sum_{j\in C_i} f_j^P(\omega) = p_i^D(\omega) && (5.9b)\\
&f_i^Q(\omega) + q_i^G(\omega) - \sum_{j\in C_i} f_j^Q(\omega) = q_i^D(\omega) && (5.9c)\\
&u_k + 2\left(r_k f_k^P + x_k f_k^Q\right) = u_{A_k} && (5.9d)\\
&\sum_{i\in\mathcal{G}} \alpha_i = 1\% && (5.9e)\\
&\sum_{i\in\mathcal{G}} \alpha_i = 1\mathrm{Cr}\left[u_k(\omega) \le u_k^{\max}\right] \ge 1 - \epsilon_v \mathrm{v} && (5.9f)\\
&\mathrm{Cr}\left[-u_k(\omega) \le -u_k^{\min}\right] \ge 1 - \epsilon_v && (5.9g)\\
&\mathrm{Cr}\left[p_i^G(\omega) \le p_i^{G,\max}\right] \ge 1 - \epsilon_g && (5.9h)\\
&\mathrm{Cr}\left[-p_i^G(\omega) \le -p_i^{G,\min}\right] \ge 1 - \epsilon_g && (5.9j)\\
&\mathrm{Cr}\left[q_i^G(\omega) \le q_i^{G,\max}\right] \ge 1 - \epsilon_g && (5.9k)\\
&\mathrm{Cr}\left[-q_i^G(\omega) \le -q_i^{G,\max}\right] \ge 1 - \epsilon_g && (5.9l)\\
&(f_i^P)^2 + (f_i^Q)^2 \le \left(S_i^{\max}\right)^2 && (5.9z)
\end{aligned}
\tag{5.9}
$$

Equation (5.9a) defines the objective function. Equations (5.9b) through (5.9d) provide linear expressions of the LinDistFlow [32] under forecast uncertainty. Specifically, Eqs. (5.9b) and (5.9c) ensure that node power balances accommodate any realization of uncertainty, while Eq. (5.9d) calculates the squared voltage magnitude at node k. Equations (5.9f) through (5.9k) represent the fuzzy chance constraints for controllable resources, ensuring that power generation and squared voltage magnitudes do not exceed their respective limits under the confidence level specified by the DSO. The expressions for $p_i^G(\omega)$ and $q_i^G(\omega)$ are provided in Eq. (5.5), with the explicit expressions for $u_i(\omega)$ detailed in Sect. 5.3. $(1 - \epsilon_g)$ and $(1 - \epsilon_v)$ denote the confidence levels of the chance constraints, while $\epsilon_g, \epsilon_v \in [0, 1/2]$ are the corresponding parameters. Finally, Eq. (5.9) represents the line capacity constraints.

5.4 Solving Strategy

To solve the FCC-ACOPF problem, two primary approaches are considered. The first approach employs a hybrid algorithm combining evolutionary techniques and fuzzy simulation. However, this method has been shown to be relatively slow and often provides approximate solutions [31]. The second approach involves converting fuzzy chance constraints into clear equivalence classes and using a solver to obtain the optimal solution. This approach offers higher solving speed and precision, but it requires that the chance constraints in the FCC-ACOPF be transformed into a specific expression [31], which can be challenging. Therefore, to achieve an accurate solution, it is crucial to convert the constraints into the required specific form to establish clear equivalence classes.

The specific expression of the fuzzy chance constraints required for this conversion can be written into [31]:

$$\begin{aligned} &\mathrm{Cr}\{\xi | g(\mathbf{x}, \boldsymbol{\xi}) \leqslant 0\} \geqslant \beta \Leftrightarrow \\ &\mathrm{Cr}\{\xi | h_0(\mathbf{x}) + h_1(\mathbf{x})\xi_1 + \cdots + h_n(\mathbf{x})\xi_n \leqslant 0\} \geqslant \beta \end{aligned} \tag{5.10}$$

where $g(\mathbf{x}, \boldsymbol{\xi})$ represents a constraint expression, $\mathbf{x}$ denotes the vector of decision variables, $\boldsymbol{\xi} = (\xi_1, ..., \xi_n)$ indicates a n-dimensional trapezoidal fuzzy vector, and $\beta = 1 - \epsilon$ refers to the confidence level. $h_i(\mathbf{x})$, where $i \in \{0, 1, ..., n\}$, stands for a function associated with the decision vector $\mathbf{x}$.

5.4.1 Fuzzy Chance Constraint of Controllable Generator

Except for the root node, it is assumed that all nodes are equipped with RES generators, i.e., $\omega_0 = 0$. Consequently, for a radial system with $n + 1$ nodes, the fuzzy vector $\boldsymbol{\omega}$ employed to express the forecast error is n-dimensional. Ω^P and Ω^Q denote the total mismatches of active and reactive power in the distribution network. Given that the forecast error fuzzy variables are independent, Ω^P and Ω^Q can be formulated as:

$$\Omega^P = \left(R_1^P, R_2^P, R_3^P, R_4^P\right), \quad \Omega^Q = \left(R_1^Q, R_2^Q, R_3^Q, R_4^Q\right) \tag{5.11}$$

where $R_m^P = \sum_{k \in \mathcal{N}^+} r_{km}$, $m \in \{1, 2, 3, 4\}$, and Ω^Q is defined by analogy.

To calculate the clear equivalence class for (5.9h), the output of the controllable generator can be expressed as follows:

$$p_i^G(\omega) = p_i^G + \alpha_i \Omega^P = p_i^G + \alpha_i \omega_1 + \alpha_i \omega_2 + \cdots + \alpha_i \omega_n \tag{5.12}$$

where $\alpha_i \geq 0$, $h_{ik}^+ = \alpha_i$, $h_{ik}^- = 0$, $i \in \mathcal{G}$, $k \in \mathcal{N}^+$. Equation (5.9h) can be rewritten as the following form:

$$\begin{aligned} &2\epsilon_g \sum_{k=1}^{n} \left(r_{k3} h_{ik}^+ - r_{k2} h_{ik}^- + (1 - 2\epsilon_g\right) \sum_{k=1}^{n} \left(r_{k4} h_{ik}^+ - r_{k1} h_{ik}^-\right) + p_i^G \leq p_i^{G,\max} \\ &\Rightarrow p_i^G + \alpha_i\left[\left(2\epsilon_g\right)R_3^P + \left(1 - 2\epsilon_g\right)R_4^P\right] \leq p_i^{G,\max} \end{aligned} \tag{5.13}$$

Similarly, the clear equivalence classes of (5.9i)–(5.9k) can be obtained.

5.4.2 Real-Time Power Flow Balance Constraint

Given the use of the affine control strategy and the predetermined participation factors, and based on the findings from [29], when the total participation factor sum is 1, the generator output, including the uncertain component, exactly matches the total demand. Consequently, (5.9b) can be reformulated into the following deterministic expression:

$$\begin{aligned} & f_i^P(\omega) + p_i^G(\omega) - \sum_{j \in C_i} f_j^P(\omega) = p_i^D(\omega) \\ & \Rightarrow p_i^D - p_i^G - f_i^P + \sum_{j = C_i} f_j^P = 0 \end{aligned} \tag{5.14}$$

Similar to (5.14), (5.9c) can also be reformulated as deterministic equality constraint.

5.4.3 Fuzzy Chance Constraint of Node Voltage

To derive the deterministic equivalence of (5.9f) and (5.9g), we follow the approach outlined in [22], which is expressed as follows:

$$u_k(\omega) = u_k - 2\big[\big(A^T RA\big)_{k*}\big(I - \alpha e^T + A^T XA\big)_{k*}\big(I - \alpha e^T\big)\mathrm{diag}(\gamma_k)\big]\boldsymbol{\omega} \tag{5.15}$$

where $e := n * 1$ represents the unit vector, and $I := n * n$ indicates identity matrix. $\gamma_k = \sqrt{1 - \cos^2 \varphi_k / cos\varphi_k}$. $R, X := l * l$ denotes diagonal matrix associated with the resistance and reactance of each line. $A := l * n$ can be expressed as:

$$a_{lk} := \begin{cases} 1, & \text{if line } l \text{ is part of the path} \\ & \text{from root to node} \\ 0, & \text{otherwise} \end{cases} \qquad l \in \mathcal{L}, k \in \mathcal{N}^+$$

For convenience, define the matrix $H := n * n$:

$$H_{k^*} = -2\big[\big(A^T RA\big)_{k*}\big(I - \alpha e^T + (A^T XA\big)_{k*}\big(I - \alpha e^T\big)\mathrm{diag}(\gamma_k)\big] \tag{5.16}$$

The node voltage amplitude squared with uncertain RESs injection can be written into:

$$u_k(\omega) = u_k + H_{k1}\omega_1 + H_{k2}\omega_2 + \cdots + H_{kn}\omega_n \tag{5.17}$$

Then, define $h_{kt}^+ = H_{kt} \vee 0$, $h_{kt}^- = -H_{kt} \wedge 0$, $t \in \mathcal{N}^+$. Thus, the chance constraint shown in (5.9f) can be relaxed into the following deterministic constraint:

$$\mathrm{Cr}\left[u_k + H_{k1}\omega_1 + \cdots + H_{kn}\omega_n - u_k^{\max} \le 0\right] \ge 1 - \epsilon_v$$
$$\Rightarrow u_k + (2\epsilon_v) \sum_{t\in\mathcal{N}^+} \left(r_{t3}h_{kt}^{+} - r_{t2}h_{kt}^{-}\right) + (1 - 2\epsilon_v) \sum_{t\in\mathcal{N}^+} \left(r_{t4}h_{kt}^{+} - r_{t1}h_{kt}^{-}\right) \le u_k^{\max} \tag{5.18}$$

However, constraint (5.18) denotes a piecewise function influenced by h_{kt}^{+} and h_{kt}^{-}, rendering the model unsolvable directly using standard solvers. To address this, this chapter linearizes the piecewise constraint (5.18) by introducing auxiliary variables, enabling the model to be solved with existing commercial solvers.

We set positive continuous variables $h1_{kt}$, $h2_{kt}$, $h3_{kt}$, $K1_{kt}$, $K2_{kt}$, binary variables $z1_{kt}$, $z2_{kt}$, and a sufficiently large positive integer M, which meet the constraints given below:

$$\begin{cases} h1_{kt} + h2_{kt} + h3_{kt} = 1 & h3_{kt} \le z2_{kt} \\ z1_{kt} + z2_{kt} = 1 & H_{kt} = -Mh1_{kt} + Mh3_{kt} \\ h1_{kt} \le z1_{kt} & K1_{kt} = Mh3_{kt} \\ h2_{kt} \le z1_{kt} + z2_{kt} & K2_{kt} = Mh1_{kt} \end{cases} \tag{5.19}$$

Therefore, Eq. (5.19) can be rewritten into the following linear expression:

$$u_k + (2\epsilon_v) \sum_{t\in\mathcal{N}^+} [r_{t3}K1_{kt} - r_{t2}K2_{kt}] + (1 - 2\epsilon_v)$$
$$\sum_{t\in\mathcal{N}^+} [r_{t4}K1_{kt} - r_{t1}K2_{kt}(x)] \le u_k^{\max} \tag{5.20}$$

Similarly to (5.20), Eq. (5.9g) can also be rewritten into a linear deterministic expression. Consequently, the chance constraints (5.9f) and (5.9g) can now be presented into the deterministic equivalence of the FCC-ACOPF:

$$\min \sum_{i=\mathcal{G}} \mathbb{E}\left[c_i\left(p_i^G(\boldsymbol{\omega})\right)\right]$$
$$\text{s.t.} \quad (8d), (8k), (12), (13), (19)$$
$$q_i^D - q_i^G - f_i^Q + \sum_{j=\mathcal{C}_i} f_j^Q = 0 - u_k + (2\epsilon_v)$$
$$\sum_{t\in\mathcal{N}^+} [r_{t3}K2_{kt} - r_{t2}K1_{kt}] + (1 - 2\epsilon_v) \sum_{t\in\mathcal{N}^+} [r_{t4}K2_{kt} - r_{t1}K1_{kt}] \le -u_k^{\min}$$
$$-p_i^G - \alpha_i\left[(2\epsilon_g)R_2^P + (1 - 2\epsilon_g)R_1^P\right] \le -p_i^{G,\min}$$
$$q_i^G + \alpha_i\left[(2\epsilon_g)R_3^Q + (1 - 2\epsilon_g)R_4^Q\right] \le q_i^{G,\max}$$
$$-q_i^G - \alpha_i\left[(2\epsilon_g)R_2^Q + (1 - 2\epsilon_g)R_1^Q\right] \le -q_i^{G,\min} \tag{5.21}$$

5.5 Case Study

To validate the FCC-ACOPF, simulation studies are performed using a modified 15-node radial system [22], which consists of controllable DERs at nodes 6 and 11. The system's topology and parameters are depicted in Fig. 5.3. The generation cost coefficients for the DERs at nodes 6 and 11 are set to $c_{0,i} = 0$, $c_{1,i} = 10\$/\text{MWh}$, and $c_{2,i} = 5\$/\text{MWh}^2$, and for the substation, the coefficients are $c_{0,i} = 0$, $c_{1,i} = 50\$/\text{MWh}$, and $c_{2,i} = 5\$/\text{MWh}^2$. These parameters are chosen to prioritize the use of controllable DERs to mitigate uncertainty in node injection power. The scheduled net load data, as shown in Table 5.1, is used with an asymmetric forecast error represented by ω_k. Here $r_{km} = a_{km}\sigma p_k^D$, $m \in \{1, 2, 3, 4\}$. σ denotes the degree of deviation of the forecast value from the actual value, where a higher σ indicates a larger potential forecast error. The four fuzzy parameters $(r_{k1}, r_{k2}, r_{k3}, r_{k4})$ of the forecast error are asymmetrical. The model is implemented using the Julia JuMP package [33].

(1) This chapter compares the performance of the proposed FCC-ACOPF model against both GCC-ACOPF and deterministic OPF (DET-ACOPF) models. In DET-ACOPF, constraints (5.9b)–(5.9k) are treated as deterministic, i.e., $\alpha_i = 0$, $\omega_i = 0$. In GCC-ACOPF, the impact of uncertainty on voltage is ignored, making constraints (5.9f)–(5.9g) deterministic.
(2) Monte Carlo sampling is used to sample the asymmetric fuzzy distribution with 1000 times. Based on these samples, the chapter assesses whether the actual frequency of node voltage violations aligns with the specified confidence level.
(3) To evaluate the feasibility of the FCC-ACOPF model for asymmetric forecast errors, this chapter compares node voltage violations and expected costs.

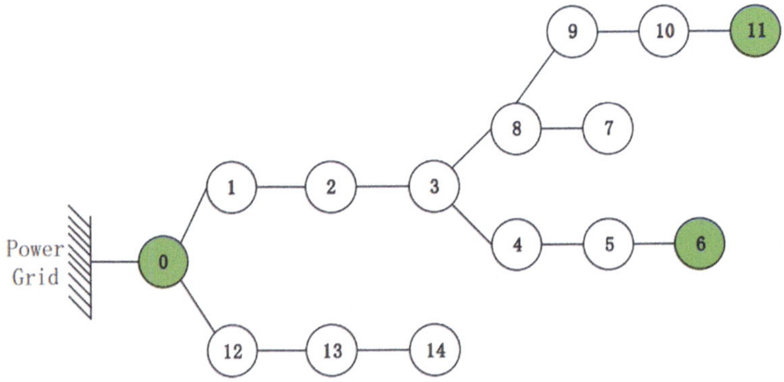

Fig. 5.3 The topology of the modified 15-node distribution system. Reprinted from Ref. [35]. Copyright 2024, with permission Electric Power Systems Research

Table 5.1 The parameters of the 15-node system

i	R_i	X_i	S_i	pD	qD	$B_i \cdot 10^{-3}$
0	–	–	–	–	–	0
1	0.001	0.12	10	0.7936	0.1855	1.1
2	0.0883	0.1262	10	0	0	2.8
3	0.1384	0.1978	10	0.0201	0.0084	2.4
4	0.0191	0.0273	10	0.0173	0.0043	0.4
5	0.0175	0.0251	10	0.0291	0.0073	0.8
6	0.0482	0.0689	10	0.0219	0.0055	0.6
7	0.0523	0.0747	10	– 0.1969	0.0019	0.6
8	0.0407	0.0582	10	0.0235	0.0059	1.2
9	0.01	0.0143	10	0.0229	0.0142	0.4
10	0.0241	0.0345	10	0.0217	0.0065	0.4
11	0.0103	0.0148	10	0.0132	0.0033	0.1
12	0.001	0.12	10	0.6219	0.1291	0.1
13	0.1559	0.1119	10	0.0014	0.0008	0.2
14	0.0953	0.0684	10	0.0224	0.0083	0.1

Reprinted from Ref. [35]. Copyright 2024, with permission Electric Power Systems Research

5.5.1 Compare Benefits of FCC-ACOPF to Other Models

To validate the proposed FCC-ACOPF model, Fig. 5.4 presents a comparison of voltage amplitudes obtained using the FCC-ACOPF model (with varying values of ϵ_v), the GCC-ACOPF model ($\epsilon_g = 0.05$), and the DET-ACOPF model in the radial system. The figure reveals a notable discrepancy between the voltage amplitudes of nodes 2 through 11 in GCC-ACOPF and DET-ACOPF compared to FCC-ACOPF. Specifically, in DET-ACOPF and GCC-ACOPF, the voltage amplitudes for nodes 2 through 11 are significantly higher and close to the system's upper voltage limit, suggesting a tendency toward overvoltage under large forecast errors. This occurs because both GCC-ACOPF and DET-ACOPF use deterministic voltage constraints that do not account for the impact of uncertainty.

In the FCC-ACOPF, voltage amplitudes are regulated by chance constraints (5.9f)–(5.9g), which ensure that voltage violations remains below a predefined threshold ϵ_v during real-time operation. As shown in Fig. 5.4, decreasing the value of ϵ_v leads to reduced voltage amplitudes for nodes 2–11, thereby increasing the voltage security margin under uncertain conditions. In contrast, node 1, which is closer to the system's lower voltage limit, is at greater risk of undervoltage in the presence of negative forecast errors. Therefore, a higher ϵ_v value in the FCC-ACOPF model leads to a decrease in the voltage amplitude of node 1.

Table 5.2 presents the participation factors for FCC-ACOPF and GCC-ACOPF under varying confidence levels ϵ. The factor functions to allocate the total mismatch

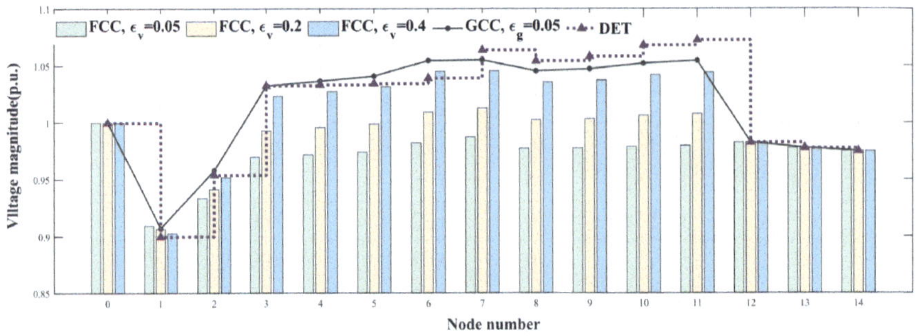

Fig. 5.4 Voltage amplitude comparison among DET, GCC and FCC-ACOPF

power caused by forecast errors to maintain demand and supply balance. As indicated in the table, the substation's participation factor in FCC-ACOPF is notably higher than in GCC-ACOPF across different confidence levels. This disparity arises because FCC-ACOPF's participation factor is constrained with (5.9h)–(5.9i). This implies that, in the presence of forecast errors, the substation must supply more power to ensure system balance and keep voltage within a safe margin.

Moreover, although the participation factors for GCC-ACOPF and FCC-ACOPF differ, they both exhibit a similar trend, which is reflected in the expected cost shown in Fig. 5.5. As ϵ increases, the substation's participation factor decreases while that of the controllable generators increases. This trend suggests that, with higher allowable risk, the system aims to maximize economic benefits by reducing the active output from the substation and relying more on controllable generators to manage the uncertainties within the distribution system.

Figure 5.5 depicts the optimal solutions for FCC-ACOPF, GCC-ACOPF, and DET-ACOPF in terms of expected generation costs, varying with different confidence levels ϵ and forecast errors σ. The expected cost, derived from Eq. (5.9a), reflects the hypothesized operational cost prior to the realization of uncertain quantities. As anticipated, the expected generation cost rises with the inclusion of additional constraints,

Table 5.2 Participation factor

$\epsilon = \epsilon_g = \epsilon_v$	GCC-A	ACOPF	case	FCC-A	ACOPF	case
	α_0	α_6	α_{11}	α_0	α_6	α_{11}
5%	0.129	0.435	0.435	0.511	0.246	0.243
10%	0.119	0.44	0.44	0.499	0.251	0.25
15%	0.108	0.446	0.446	0.485	0.256	0.257
20%	0.09	0.453	0.453	0.468	0.266	0.266
25%	0.075	0.462	0.462	0.446	0.276	0.277
30%	0.05	0.475	0.475	0.415	0.292	0.293
35%	0.017	0.491	0.491	0.371	0.314	0.314

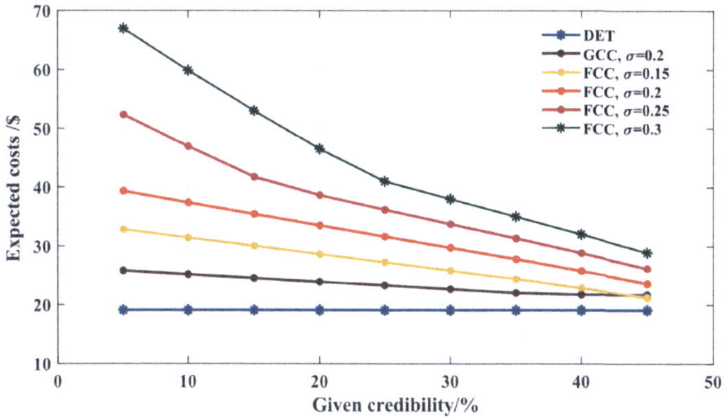

Fig. 5.5 The expected costs for different values of $\epsilon(\epsilon = \epsilon_v = \epsilon_g)$ and σ

which ensures larger security margins for the distribution system. Conversely, the value of σ represents the deviation between the forecasted and actual net load values. As illustrated in Fig. 5.5, a higher σ implies a greater potential forecast error, leading to increased costs for system balancing in real-time operation.

Additionally, ϵ denotes the risk level in the scheduling model. A lower ϵ indicates a reduced tolerance for constraint violations, necessitating more conservative and costly decision-making. In fact, cost and risk are interrelated; relaxing constraint violations can lead to higher operational costs, but excessive relaxation may result in frequent voltage limit violations. Thus, selecting an appropriate confidence level is crucial for balancing risk and cost in OPF with RES integration.

5.5.2 FCC-ACOPF Model Validation

Monte Carlo method is used to sample forecast errors and assess whether the actual frequency of node voltage violations aligns with the specified confidence level. Given that fuzzy forecast errors cannot be sampled directly, we employ the approach outlined in [33] to generate a set of potential fuzzy samples. The probability density function (PDF) of the trapezoidal fuzzy number $(r_{i1}, r_{i2}, r_{i3}, r_{i4})$ is as follows:

$$f(\omega_i) = \begin{cases} \frac{\omega_i - r_{i1}}{r_{i2} - r_{i1}} h_i, & r_{i1} \leqslant \omega_i \leqslant r_{i2} \\ h_i, & r_{i2} \leqslant \omega_i \leqslant r_{i3} \\ \frac{\omega_i - r_{i4}}{r_{i3} - r_{i4}} h_i, & r_{i3} \leqslant \omega_i \leqslant r_{i4} \\ 0, & \text{otherwise} \end{cases} \tag{5.22}$$

where $h_i = [(r_{i4} - r_{i1}) + (r_{i3} - r_{i2})]/2$ represents the calibration parameter. Note that the sum of all possibility (total area under the PDF curve) should be equal to 1.

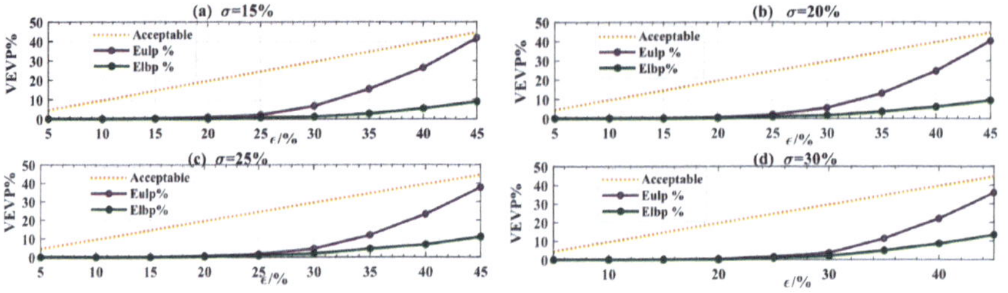

Fig. 5.6 The trend chart of voltage empirical violation frequency under different values of $\epsilon(\epsilon = \epsilon_v = \epsilon_g)$ and σ. VEVF: voltage empirical violation frequency, VULVF: voltage upper limit violation frequency, VLLVF: voltage lower limit violation frequency

Using Eq. (5.21), the inverse cumulative probability distribution is derived. Subsequently, 1000 sets of samples are generated through the Monte Carlo. Figure 5.6 illustrates the empirical violation of node voltage for different $\epsilon(\epsilon = \epsilon_v = \epsilon_g)$ and σ. With the FCC-ACOPF, the DSO mandates that the empirical violation must not exceed the predefined violation level ϵ. A reformulation leading to a high violation could result in insecure operations and potential infeasibility.

As depicted in Fig. 5.6, the empirical violation for FCC-ACOPF remains below the acceptable confidence level, and as ϵ decreases, the frequency of voltage violations also decreases. This demonstrates the model's ability to balance security and cost effectively. Furthermore, the empirical violation of FCC-ACOPF aligns with expectations across different values of σ, confirming the model's feasibility.

Figure 5.7 illustrates the voltage violation for each node under varying ϵ. As anticipated, tightening the chance constraints by reducing ϵ decreases the frequency of violations, and for some nodes, may even eliminate them entirely. This observation underscores that ϵ reflects the level of risk.

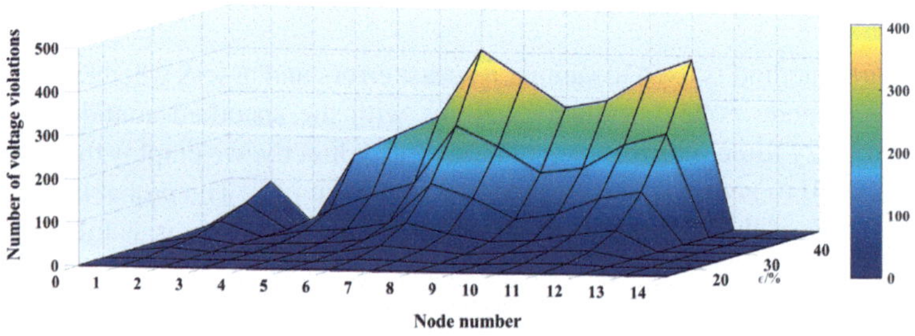

Fig. 5.7 Node voltage violation times for different values of $\epsilon(\epsilon = \epsilon_v = \epsilon_g)$

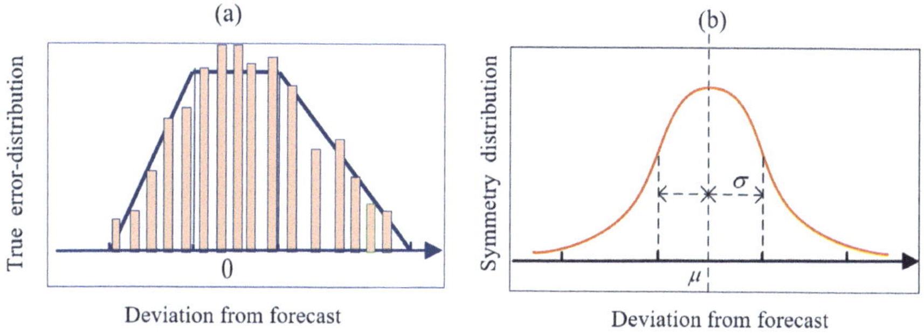

Fig. 5.8 **a** True (unknown) error distribution. **b** Symmetric normal distribution with mean and variance

5.5.3 Uncertainty Analysis of FCC-ACOPF and CC-ACOPF

Figure 5.8a displays the forecast error distribution obtained from 1000 samples, highlighting its significant asymmetry. The FCC-ACOPF model is applied using this asymmetric distribution. In contrast, Fig. 5.8b illustrates a normal distribution derived from the mean and variance of the distribution shown in Fig. 5.8a, which is used for implementing the CC-ACOPF model.

The voltage violation frequencies for the FCC-ACOPF and CC-ACOPF models, evaluated using the same set of forecast error samples from Fig. 5.8a, are compared in Fig. 5.9a. It is evident that when the error distribution for RES nodes exhibits significant asymmetry, the voltage violation for the CC-ACOPF (A), which is a normal distribution, exceeds the prescribed confidence level. This elevated violation frequency can lead to incorrect operational decisions. Conversely, the FCC-ACOPF (A) model maintains the voltage violation below the confidence level, demonstrating its greater reliability in managing the uncertainties associated with asymmetric forecast errors.

To validate that the higher voltage violations observed with the CC-ACOPF model are indeed due to asymmetry, similar analyses were performed on two additional sets of forecast error samples, labeled (B) and (C). The results show that as the error distribution becomes less asymmetric, the voltage violation for the CC-ACOPF model decreases.

Conversely, Fig. 5.9b illustrates that the expected cost of FCC-ACOPF is 47.9% higher than that of CC-ACOPF when $\epsilon = 5\%$. This increased cost for FCC-ACOPF is accompanied by a higher security margin. For instance, when the allowable violation frequency is set at 5%, the CC-ACOPF model results in a violation frequency of 11.3%, which clearly exceeds the acceptable limit. In contrast, the FCC-ACOPF model maintains a violation frequency of 0%. Moreover, as ϵ increases, the gap between the FCC-ACOPF and CC-ACOPF costs narrows. The expected cost for CC-ACOPF is lower, but it fails to ensure compliance with acceptable violation limits. Practically, DSOs tend to favor solutions that are more conservative and reliable, even if they entail higher costs.

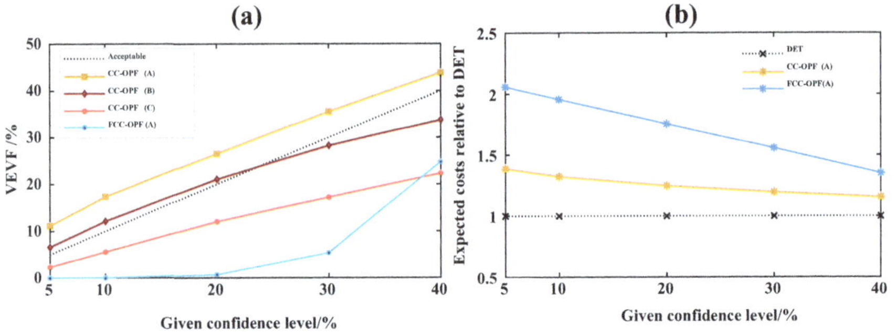

Fig. 5.9 The comparison of node voltage violations and expected costs for FCC-OPF and CC-OPF. Forecast error samples (A) obtained from the distribution of 7(a), forecast error sample (B) obtained from an asymmetric distribution, and forecast error sample obtained (C) from a symmetric distribution

5.6 Conclusion

In this chapter, we introduce credibility theory into OPF of renewable distribution networks, proposing FCC-ACOPF model specifically for radial distribution systems. This model addresses the uncertainty inherent in renewable energy sources with asymmetric features. The FCC-ACOPF model facilitates optimal scheduling under various confidence levels, enabling DSO to achieve the desired objectives by setting an acceptable confidence level. Simulation results demonstrate the following:

(1) The FCC-ACOPF model effectively meets prescribed risk criteria across different confidence levels and is computationally feasible.
(2) In evaluations involving asymmetric sample performance, FCC-ACOPF shows superior performance compared to CC-ACOPF with a moderate increase in expected cost. This advantage is due to CC-ACOPF's inability to guarantee theoretical violation limits. The FCC-ACOPF offers a novel tool and perspective for managing uncertain risks in power distribution systems.

References

1. Akbari B, Sansavini G (2023) Adaptive robust ac optimal power flow considering intrahour uncertainties. Electr Power Syst Res 216:109082
2. Lee D, Turitsyn K, Molzahn DK, Roald LA (2021) Robust ac optimal power flow with robust convex restriction. IEEE Trans Power Syst 36(6):4953–4966
3. Li H, Hou K, Xu X, Jia H, Zhu L, Mu Y (2022) Probabilistic energy flow calculation for regional integrated energy system considering cross-system failures. Appl Energy 308:118326
4. Maheshwari A, Sood YR, Jaiswal S (2023) Flow direction algorithm-based optimal power flow analysis in the presence of stochastic renewable energy sources. Electr Power Syst Res 216:109087

5. Lu R, Ding T, Qin B, Ma J, Fang X, Dong Z (2019) Multi-stage stochastic programming to joint economic dispatch for energy and reserve with uncertain renewable energy. IEEE Trans Sustain Energy 11(3):1140–1151
6. Senthilraja S, Gangadevi R, Ko¨ ten H, Marimuthu R (2020) Performance assessment of a solar powered hydrogen production system and its anfis model. Heliyon 6(10):e05271
7. Lorca A, Sun XA (2018) The adaptive robust multi-period alternating current optimal power flow problem. IEEE Trans Power Syst 33(2):1993–2003
8. Zhao P, Gu C, Huo D, Shen Y (2020) Two-stage distributionally robust optimization for energy hub systems. IEEE Trans Ind Inf 16(5):3460–3469
9. Chen J, Wu Q, Zhang L, Wu P (2017) Multi-objective mean–variance–skewness model for nonconvex and stochastic optimal power flow considering wind power and load uncertainties. Eur J Oper Res 263(2):719–732
10. Gallego LA, Franco JF, Cordero LG (2021) A fast-specialized point estimate method for the probabilistic optimal power flow in distribution systems with renewable distributed generation. Int J Electr Power Energy Syst 131:107049
11. Li S, Gong W, Wang L, Gu Q (2022) Multi-objective optimal power flow with stochastic wind and solar power. Appl Soft Comput 114:108045
12. Wang S, Zhao C, Fan L, Bo R (2022) Distributionally robust unit commitment with flexible generation resources considering renewable energy uncertainty. IEEE Trans Power Syst 37(6):4179–4190
13. Ansaripour R, Barati H, Ghasemi A (2022) A chance-constrained optimization framework for transmission congestion management and frequency regulation in the presence of wind farms and energy storage systems. Electr Power Syst Res 213:108712
14. Chen G, Zhang H, Hui H, Song Y (2022) Deep-quantile-regression-based surrogate model for joint chance- constrained optimal power flow with renewable generation. IEEE Trans Sustain Energy 14(1):657–672
15. Fathabad AM, Cheng J, Pan K, Yang B (2023) Asymptotically tight conic approximations for chance- constrained ac optimal power flow. Eur J Oper Res 305(2):738–753
16. Pena-Ordieres A, Molzahn DK, Roald LA, Wa¨chter A (2021) Dc optimal power flow with joint chance constraints. IEEE Trans Power Syst 36(1):147–58
17. Aigner KM, Clarner JP, Liers F, Martin A (2022) Robust approximation of chance constrained dc optimal power flow under decision-dependent uncertainty. Eur J Oper Res 301(1):318–333
18. Chen T, Song Y, Hill DJ, Lam AY (2022) Chance-constrained OPF in droop-controlled microgrids with power flow routers. IEEE Trans Smart Grid 13(4):2601–2613
19. Lei X, Yang Z, Zhao J, Yu J (2022) Data-driven assisted chance-constrained energy and reserve scheduling with wind curtailment. Appl Energy 321:119291
20. Lee D, Han C, Kang S, Jang G (2023) Chance-constrained optimization for active distribution networks with virtual power lines. Electr Power Syst Res 221:109449
21. Mieth R, Dvorkin Y (2020) Distribution electricity pricing under uncertainty. IEEE Trans Power Syst 35(3):2325–2338
22. Mieth R, Dvorkin Y (2018) Data-driven distributionally robust optimal power flow for distribution systems. IEEE Control Syst Lett 2(3):363–368
23. Gonzalez-Castellanos A, Hinneck A, Mieth R, Pozo D, Dvorkin Y (2021) Electricity and reserve pricing in chance-constrained electricity markets with asymmetric balancing reserve policies. arXiv preprint arXiv:210605999 2021
24. Haupt Sue Ellen C et al (2019) The use of probabilistic forecasts: applying them in iheory and practice. IEEE Power Energy Mag 17(6):46–57
25. Liu B (2006) A survey of credibility theory. Fuzzy Optim Decis Making 5(4):387–408
26. Zhou J, Li X, Pedrycz W (2016) Mean-semi-entropy models of fuzzy portfolio selection. IEEE Trans Fuzzy Syst 24(6):1627–1636
27. Liu B (2010) Uncertainty theory. Springer
28. Attaviriyanupap P, Kita H, Tanaka E, Hasegawa J (2004) A fuzzy-optimization approach to dynamic eco- nomic dispatch considering uncertainties. IEEE Trans Power Syst 19(3):1299–1307

29. Bienstock D, Chertkov M, Harnett S (2014) Chance-constrained optimal power flow: risk-aware network control under uncertainty. SIAM Rev 56(3):461–495
30. Zhou A, Yang M, Wang M, Zhang Y (2020) A linear programming approximation of distributionally ro- bust chance-constrained dispatch with wasserstein distance. IEEE Trans Power Syst 35(5):3366–3377
31. Liu B, Liu YK (2002) Expected value of fuzzy variable and fuzzy expected value models. IEEE Trans Fuzzy Syst 10(4):445–450
32. Baran ME, Wu FF (1989) Optimal capacitor placement on radial distribution systems. IEEE Trans Power Delivery 4(1):725–734
33. Dunning I, Huchette J, Lubin M (2017) Jump: a modeling language for mathematical optimization. SIAM Rev 59(2):295–320
34. Akdemir HG (2020) Simulation studies for credibility-based multi-objective linear programming problems with fuzzy parameters. TWMS J Appl Eng Math 10(2):483–491
35. Duan JH, Chen JJ, Liu FW et al (2023) Credibility theory to handle uncertain renewable energy: a fuzzy chance constrained AC optimal power flow. Electric Power Syst Res 223:109550

Chapter 6
Multi-energy Hub Optimization to Enhance Resilience of Renewable Agricultural Microgrid

6.1 Introduction

Energy supply in rural areas serves as a foundational pillar for agricultural development. The implementation of agricultural microgrids holds the potential to fully leverage the resource abundance of rural regions, facilitating local consumption of distributed renewable resources. This, in turn, addresses challenges such as inadequate power supply and unreliable power delivery [1, 2]. Moreover, the establishment of multi-energy microgrids stands to significantly enhance operational efficiency and support the integration of renewable energy sources [3, 4]. However, a notable discrepancy persists between the substantial demand potential in rural areas and the often underdeveloped state of grid infrastructure, particularly in regions prone to natural disasters. Such areas frequently grapple with the threat of supply disruptions [5]. Therefore, it becomes imperative to bolster the agricultural microgrids resilience against supply disruptions and implement measures capable of withstanding high-impact, low-probability extreme events.

The concept of an energy hub embodies the capacity to produce, convert, store, and consume various energy carriers while carefully considering the complementarity and synergies among them [6, 7]. During extreme events, an energy hub can play a crucial role in supplying loads with the assistance of energy storage, ensuring the secure and stable operation of agricultural microgrids. For instance, in [8], a scheduling model for an energy hub based on cooling, heating, and power systems is proposed. In another study [9], the incorporation of a water-based energy storage unit into an energy hub is explored to enhance operational performance, resulting in significant reductions in pollution emissions and improvements in power supply reliability.

J. Chen and Y. Li, *Renewable Power System Optimization*, Smart Energy Systems,
https://doi.org/10.1007/978-981-97-8132-4_6

Additionally, research by refs. [10] and [11] delves into the impact of carbon mechanisms on the economic and environmental facets of multi-energy systems, fostering reductions in carbon emissions. However, the existing literature does not directly address agricultural microgrids, particularly regarding their specific requirements related to power, thermal energy, and irrigation systems. Furthermore, scant attention has been given to the investigation of electricity-heat-water hubs in enhancing agricultural microgrids resilience, especially in scenarios where water supply proves insufficient. This highlights a notable gap in the current body of research, signaling the need for further exploration and tailored solutions to bolster the resilience and efficiency of agricultural microgrid systems.

In the context of enhancing the resilience of agricultural microgrids, it is crucial to consider the issue of operational costs. For instance, in [12], metrics for evaluating flexibility are established to enable electrical and thermal supply to microgrids. Meanwhile, [13] devises energy management tailored for pumped storage systems, aiming to achieve optimal irrigation and pumping strategies, along with calculating costs and payback periods. However, these references primarily focus on economic considerations and do not specifically deal with the resilience enhancement issue.

To address this critical aspect, [14] explores a resilience-oriented proactive methodology aimed at bolstering resistance against natural hazards. This methodology emphasizes the balance between preparation indices and load supply. Additionally, [15] develops methods for vulnerability and reliability assessments in multi-energy microgrids, leveraging energy hub modeling. Nonetheless, this work does not delve into the impact of these strategies on operational costs. Additionally, these studies mentioned above often tend to treat resilience and operational costs as separate objectives, lacking an integrated consideration of these crucial facets.

Furthermore, the irradiation intensity crucial for photovoltaic generation is inherently random and intermittent, with prediction errors that cannot be entirely eliminated [16, 17]. Clearly, with the integration of photovoltaic systems, the uncertain impacts on agricultural microgrids must be carefully considered. While embedded multi-energy complementarity has shown promise in strengthening the resilience of weaker agricultural microgrids, it is noteworthy that the electricity uncertainties stemming from photovoltaic can be transferred to other energy supply parts, potentially leading to operational instabilities [18].

To mitigate such instabilities, existing literature has proposed three primary techniques: robust, fuzzy, and stochastic optimizations. For instance, in [19], a robust optimization model is employed to represent uncertain parameters as deterministic parameters, achieving a balance between cost efficiency and system robustness. Similarly, [20] introduces an optimal fuzzy logic to manage uncertainties. Additionally, [21] focuses on optimizing the renewable energy outputs through scene-based approach, with simulation results confirming the efficacy of the proposed scheduling approach in reducing the operational costs of grid-connected microgrids.

Moreover, in the context of agricultural microgrids, the utilization of both photovoltaic and solar thermal can significantly enhance the synergistic potential, leveraging the advantages of multi-energy complementarity [22]. However, it is undeniable that the random and intermittent nature of irradiation intensity not only affects solar thermal output but also contributes to instabilities in photovoltaic and solar thermal agricultural microgrids.

Despite these challenges, the existing literature on multi-energy agricultural microgrids has notably few works dedicated specifically to addressing uncertainties related to photovoltaic and solar thermal technologies. This presents an opportunity for further research and exploration into effective strategies to manage and optimize the performance of these systems amidst unpredictable irradiation intensity. By filling this gap, future studies can contribute significantly to the development of resilient and efficient multi-energy agricultural microgrid systems, ensuring sustainable energy supply for rural communities while minimizing operational risks and costs.

In previous works, various optimization algorithms have been employed to tackle multi-objective optimization challenges in microgrid scheduling. For instance, in [23], a particle swarm optimization was utilized to address multi-objective optimization. Meanwhile, [24] introduced a multi-objective group search optimization to optimize grid-connected microgrid, focusing on objectives such as operational costs and pollution emissions. Additionally, [25] developed a multi-objective self-adaptive differential evolution tailored for energy storage planning in microgrids.

Despite these advancements, existing algorithms may struggle to strike a balance between local exploitation and global exploration capabilities, particularly when confronted with the complexities inherent in power system optimization. Such challenges can result in these algorithms converging to local optima rather than achieving a globally optimal solution.

To underscore the innovation of our work, a comprehensive review and comparison of literature focusing on the resilience enhancement of agricultural microgrids have been investigated. Through an analysis of the table and existing literature, several key observations emerge: (1) A noticeable research gap exists as few works have focused on implementing an electricity-heat-water hub to enhance the resilience of electricity, heat, and irrigation simultaneously within agricultural microgrids. (2) Despite the exploration of photovoltaic power generation utilization and consumption in various studies, there is a lack of research addressing the impacts of uncertainties associated with photovoltaic and solar thermal technologies. (3) Existing literature predominantly concentrated on optimizing individual objective related to either resilience or cost, often leading to conflicting objectives between economic considerations and resilience goals.

In light of these identified research gaps, this chapter contributes significantly to the field in the following ways:

(1) An innovative concept of electricity-heat-water-based multi-energy hub (EHWbMEH) tailored for agricultural microgrids is proposed. This conceptualization is motivated by the overarching objective of enhancing the reliability of energy supply, particularly during adverse events.
(2) The stochastic characteristics and coupling relationships between photovoltaic and solar thermal are comprehensively investigated by introducing Expected Power Not Served (EPNS) and Expected Power Curtailment (EPC).
(3) A novel EHWbMEH-based multi-objective optimization is presented to address resilience improvement of agricultural microgrids. To solve this intricate model, a Multiple Preys-based Evolutionary Predator and Prey Strategy (MPEPPS) is employed.

The structure of the chapter is structured as follows: Sect. 6.2 presents the concept of EHWbMEH. Then, the multi-objective resilience-economic optimization for agricultural microgrids and details the proposed MPEPPS are given by Sect. 6.3. Section 6.4 provides insights from simulations conducted on the EHWbMEH-based agricultural microgrid. Finally, conclusions and key findings are summarized in Sect. 6.5.

6.2 Electricity–Heat–Water-Based Multi-energy Hub

Figure 6.1 provides the structural of EHWbMEH system. It integrates several components to supply electricity, heat, and water, including a photovoltaic system, a solar thermal system, a pumped storage system, a combined heat and power (CHP) unit, a thermal storage unit (TSU), and a heat pump (HP). The upstream grid and natural gas provide the primary energy inputs, while the photovoltaic and solar thermal systems contribute electricity and heat to the EHWbMEH system. The operational models for the photovoltaic and solar thermal systems, the pumped storage system, the HP, the CHP, and the TSU are detailed as follows.

6.2.1 *Photovoltaic and Solar Thermal System*

Photovoltaic panels convert solar irradiation into power. These panels' outputs are influenced by the intensity of the irradiation and the ambient temperature, as described in [27]:

$$P_{\mathrm{PV}}^{t} = 10^{-6}\mu_{\mathrm{cell}}S_{\mathrm{PV}}[1 - 0.005(T_{\mathrm{e}} - 25)]E^{t} \tag{6.1}$$

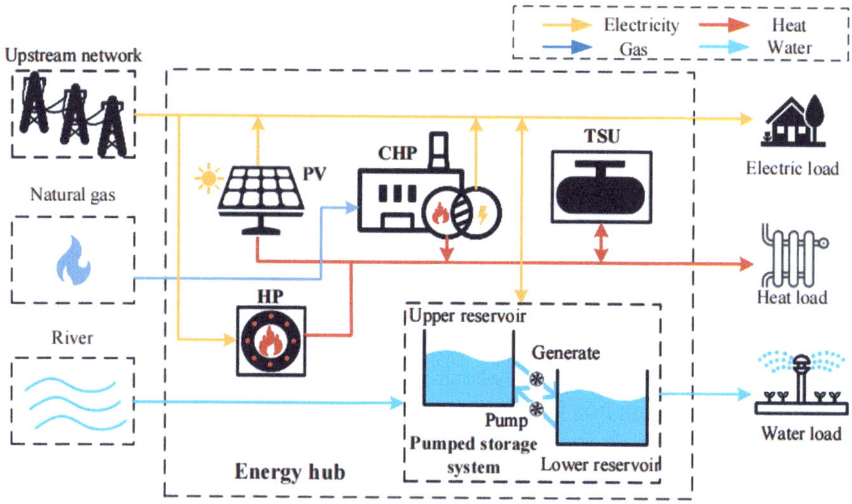

Fig. 6.1 The structure of the EHWbMEH, reprinted from Ref. [26]. Copyright 2024, with permission renewable energy

where P^t_{PV} represents photovoltaic panels output; μ_{cell} denotes the conversion efficiency (irradiation intensity is 1000 W/m^2, atmospheric mass is 1.5 and cell temperature is 25 °C); S_{PV} denotes photovoltaic panels area; T_{e} indicates the cell temperature; E^t refers to the irradiation intensity at time t.

The heat absorbing from solar irradiation is:

$$H^t_{\text{PV}} = 10^{-6}\mu_{\text{SC}} S_{\text{PV}} E^t \tag{6.2}$$

where H^t_{PV} denotes the heat from solar collector and μ_{SC} indicates the solar thermal conversion efficiency.

Let H^t_{sp} represent the heat delivered to the power, and H^t_{st} denote the heat supplied to the thermal system, satisfying $H^t_{\text{PV}} = H^t_{\text{sp}} + H^t_{\text{st}}$. The electricity generated by the power P^t_{sp} with conversion efficiency η_{sp} can be calculated as $P^t_{\text{sp}} = \eta_{\text{sp}} H^t_{\text{sp}}$.

6.2.2 Pumped Storage System

The pumped storage system plays a crucial role in mitigating the variability of renewable energy sources within a multi-energy agricultural microgrid [28]. This system operates by storing surplus energy as gravitational potential energy during periods of low electricity prices. Water is pumped to an upper reservoir, and this stored energy can later be converted into clean electricity by releasing the water through a turbine into a lower reservoir when needed. In this setup, the upper reservoir is replenished by a river, while the lower reservoir is linked to the irrigation system to supply water to farmland, which is described as follows:

$$\begin{cases} P_{\mathrm{PS,P}}^{t} = a_{\mathrm{PS}}\left(Q_{\mathrm{P}}^{t}\right)^{2} + b_{\mathrm{PS}}Q_{\mathrm{P}}^{t} + c_{\mathrm{PS}} \\ P_{\mathrm{PS,G}}^{t} = a_{\mathrm{PS}}\left(Q_{\mathrm{G}}^{t}\right)^{2} + b_{\mathrm{PS}}Q_{\mathrm{G}}^{t} + c_{\mathrm{PS}} \end{cases} \tag{6.3}$$

$$\begin{cases} V_{\mathrm{up}}^{t} = V_{\mathrm{up}}^{t-1} - Q_{\mathrm{G}}^{t-1} + Q_{\mathrm{P}}^{t-1} + Q_{\mathrm{Riv}}^{t} \\ V_{\mathrm{dn}}^{t} = V_{\mathrm{dn}}^{t-1} + Q_{\mathrm{G}}^{t-1} - Q_{\mathrm{P}}^{t-1} - Q_{\mathrm{Ir}}^{t-1} \end{cases} \tag{6.4}$$

Equation (6.3) describes the hydro-electric conversion for water pumping and releasing of the pumped storage system, where $P_{\mathrm{PS,P}}^{t}$ and $P_{\mathrm{PS,G}}^{t}$ denote the power generation and consumption by releasing and pumping water at time t, respectively; a_{PS}, b_{PS} and c_{PS} are conversion coefficients. Equation (6.4) captures the change in water volume in both the upper and lower reservoirs at time t, where Q_{Riv}^{t} signifies the water volume flowing into the upper reservoir and Q_{Ir}^{t-1} represents the water volume supplied to the irrigation system.

Additionally, the reservoirs must adhere to both upper and lower capacity constraints. Furthermore, the flow rates for pumping and releasing water should be confined within specified minimum and maximum limits. These constraints can be formulated as follows:

$$V_{\mathrm{up}}^{\min} \leq V_{\mathrm{up}}^{t} \leq V_{\mathrm{up}}^{\max}, V_{\mathrm{dn}}^{\min} \leq V_{\mathrm{dn}}^{t} \leq V_{\mathrm{dn}}^{\max} \tag{6.5}$$

$$Q_{\mathrm{G}}^{\min} \leq Q_{\mathrm{G}}^{t} \leq Q_{\mathrm{G}}^{\max}, Q_{\mathrm{P}}^{\min} \leq Q_{\mathrm{P}}^{t} \leq Q_{\mathrm{P}}^{\max} \tag{6.6}$$

where $V_{\mathrm{up}}^{\min}$ and $V_{\mathrm{up}}^{\max}$ denote the minimum and maximum water volumes that can be stored in the upper reservoir, respectively; $V_{\mathrm{dn}}^{\min}$ and $V_{\mathrm{dn}}^{\max}$ represent the minimum and maximum water volumes in the lower reservoir, respectively; $Q_{\mathrm{G}}^{\min}$ and $Q_{\mathrm{G}}^{\max}$ denote the minimum and maximum flow rates for releasing water from the upper reservoir, respectively; $Q_{\mathrm{P}}^{\min}$ and $Q_{\mathrm{P}}^{\max}$ are the minimum and maximum flow rates for pumping water into the lower reservoir, respectively.

6.2.3 Combined Heat and Power Unit

The CHP unit uses natural gas to simultaneously produce heat and electricity. The generated electricity can either be used to power the pumped storage system or be consumed by the HP unit, while the heat can be stored in the TSU or used to meet the thermal load. The model for the CHP is expressed as follows:

$$P_{\mathrm{CHP}}^{t} = \eta_{\mathrm{e}} Q_{\mathrm{gas}}^{t} \tag{6.7}$$

$$H_{\mathrm{CHP}}^{t} = \eta_{\mathrm{h}} Q_{\mathrm{gas}}^{t} \tag{6.8}$$

$$Q_{\mathrm{gas}}^{\min} \leq Q_{\mathrm{gas}}^{t} \leq Q_{\mathrm{gas}}^{\max} \tag{6.9}$$

where P_{CHP}^{t} and η_{e} indicate the electricity generation and conversion coefficient from gas to electricity, respectively; H_{CHP}^{t} and η_{h} denote the heat generation and the gas-to-heat conversion coefficient. Equation (6.9) specifies the amount of natural gas consumed by CHP and adheres to the minimum and maximum constraints.

6.2.4 *Heat Pump Unit*

The HP is designed to produce substantial amounts of heat with minimal power consumption, thereby significantly reducing primary energy use. This chapter's proposed HP unit provides additional heat to the load when the CHP unit's heat supply is insufficient. The heat generated by the HP unit during hour t, denoted as H_{HP}^{t}, can be calculated using the following equation:

$$H_{\mathrm{HP}}^{t} = \eta_{\mathrm{HP}} P_{\mathrm{HP}}^{t} \tag{6.10}$$

where η_{HP} represents the heat production efficiency of the HP, and P_{HP}^{t} denotes the electricity consumed by the HP at time t. The electricity consumption of the HP unit is constrained by the following condition:

$$P_{\mathrm{HP}}^{\min} \leq P_{\mathrm{HP}}^{t} \leq P_{\mathrm{HP}}^{\max} \tag{6.11}$$

where $P_{\mathrm{HP}}^{\min}$ and $P_{\mathrm{HP}}^{\max}$ represent the minimum and maximum limits of HP, respectively.

6.2.5 *Thermal Storage Unit*

The TSU is employed to store excess energy from the CHP unit or provide additional heat when the heat supply is inadequate. By incorporating the TSU into the EHWbMEH, peak shaving and valley filling are facilitated, thereby reducing the operational costs. The model for the TSU is expressed as follows:

$$H_{\mathrm{TSU}}^{t} = H_{\mathrm{TSU}}^{t-1} + H_{\mathrm{ch}}^{t-1} \tag{6.12}$$

$$-H_{\mathrm{ch}}^{\max} \leq H_{\mathrm{ch}}^{t} \leq H_{\mathrm{ch}}^{\max} \tag{6.13}$$

$$H_{\mathrm{TSU}}^{\min} \leq H_{\mathrm{TSU}}^{t} \leq H_{\mathrm{TSU}}^{\max} \tag{6.14}$$

where H_{ch}^{t-1} denotes the charging or discharging heat of TSU at time $t-1$. The bounds of H_{ch}^{t} and H_{TSU}^{t} are limited in Eqs. (6.13) and (6.14), where $H_{\mathrm{ch}}^{\max}$ denotes the maximum of TSU charging and discharging heat; $H_{\mathrm{TSU}}^{\min}$ and $H_{\mathrm{TSU}}^{\max}$ indicate the minimum and maximum heat volume stored in TSU, respectively.

The power generated by the EHWbMEH system is used to meet both the electricity load and the needs of the irrigation system. Additionally, irrigation schedules and quantities can be adjusted based on the status of the agricultural microgrid, which can be expressed as follows:

$$P_{\mathrm{Ir}}^{t} = a_{\mathrm{Ir}}(Q_{\mathrm{Ir}}^{t})^{2} + b_{\mathrm{Ir}}Q_{\mathrm{Ir}}^{t} + c_{\mathrm{Ir}} \tag{6.15}$$

$$\sum_{t=1}^{T} Q_{\mathrm{Ir}}^{t} = Q_{\mathrm{TW}} \tag{6.16}$$

$$Q_{\mathrm{Ir}}^{\min} \leq Q_{\mathrm{Ir}}^{t} \leq Q_{\mathrm{Ir}}^{\max} \tag{6.17}$$

where Q_{Ir}^{t} denotes the water consumption of the irrigation system at time t; a_{Ir}, b_{Ir} and c_{Ir} indicate the conversion coefficients. The total daily water demand of the irrigation system Q_{TW} is given as Eq. (6.16). The water consumption of the irrigation system Q_{Ir}^{t} satisfies the constraint shown in Eq. (6.17), where $Q_{\mathrm{Ir}}^{\max}$ and $Q_{\mathrm{Ir}}^{\min}$ denote the maximum and minimum limits of the water flow, respectively.

6.2.6 *Additional Constraints for Power Balance*

In EHWbMEH system, combined with the electricity purchased from the upstream grid, the following constraint must satisfy:

$$P_{\mathrm{PS,G}}^{t} + P_{\mathrm{PV}}^{t} + P_{\mathrm{sp}}^{t} + P_{\mathrm{N}}^{t} + P_{\mathrm{CHP}}^{t} = P_{\mathrm{PS,P}}^{t} + P_{\mathrm{Ir}}^{t} + P_{\mathrm{HP}}^{t} + P_{\mathrm{load}}^{t} \tag{6.18}$$

where P_{load}^{t} represents the electricity demand; P_{N}^{t} is the electricity purchased from the upper grid.

Additionally, let H_{load}^{t} denote the heat demand. The heat flow must satisfy:

$$H_{\mathrm{HP}}^{t} + H_{\mathrm{CHP}}^{t} + H_{\mathrm{st}}^{t} = H_{\mathrm{ch}}^{t} + H_{\mathrm{load}}^{t} \tag{6.19}$$

6.3 Multi-objective Resilience-Economic Optimization

In addressing the resilience enhancement of agricultural microgrids, this chapter introduces a multi-objective resilience-economic optimization model for an EHWbMEH system, incorporating uncertainties in photovoltaic and solar thermal resources. The model aims to determine the optimal output of each unit within the EHWbMEH system to maximize the resilience of the agricultural microgrid while minimizing operational costs. It is important to prioritize critical loads when resources are available, with the model balancing the resilience of non-critical loads against operational expenses.

The EPNS and EPC risk indices are developed to manage the impact of irradiation intensity uncertainty on agricultural microgrid. These indices help in achieving an optimal trade-off between the abandonment of photovoltaic and solar thermal generation and their effective utilization. The resilience-economic model is characterized as a nonlinear, non-convex multi-objective optimization, which will be explored in detail in this subsection.

6.3.1 Uncertainty Analysis of Renewable Energy

In the first step, the uncertainty parameters encompassing photovoltaic, solar thermal, and power generation are identified. Given that the irradiation intensity throughout the day follows a normal distribution, denoted as $E^t \sim N(\mu_{\mathrm{E}}^t, (\sigma_{\mathrm{E}}^t)^2)$, the distributions of the photovoltaic, solar thermal and power generation are also expressed accordingly $P_{\mathrm{PV}}^t \sim N(\mu_{\mathrm{PV}}^t, (\sigma_{\mathrm{PV}}^t)^2)$, $H_{\mathrm{st}}^t \sim N(\mu_{\mathrm{st}}^t, (\sigma_{\mathrm{st}}^t)^2)$ and $P_{\mathrm{sp}}^t \sim N(\mu_{\mathrm{sp}}^t, (\sigma_{\mathrm{sp}}^t)^2)$. Here, μ_{E}^t, μ_{PV}^t,μ_{st}^t and μ_{sp}^t indicate the forecast values of irradiation intensity, photovoltaic generation, solar thermal generation and power generation at time t, respectively; σ_{E}^t, σ_{PV}^t, σ_{st}^t and σ_{sp}^t represent the standard deviations of irradiation intensity, photovoltaic generation, solar thermal generation and power generation at time t, respectively. Then, we assume that $H_{\mathrm{st}}^t = \alpha H_{\mathrm{PV}}^t$ and $P_{\mathrm{sp}}^t = (1-\alpha)\eta_{\mathrm{sp}} H_{\mathrm{PV}}^t$ with proportion coefficient α, so it can be deduced that:

$$\begin{cases} \mu_{\mathrm{PV}}^t = \mu_{\mathrm{cell}} S_{\mathrm{PV}}(1 - 0.005(T_{\mathrm{e}} - 25))\mu_{\mathrm{E}}^t \\ \sigma_{\mathrm{PV}}^t = \mu_{\mathrm{cell}} S_{\mathrm{PV}}(1 - 0.005(T_{\mathrm{e}} - 25))\sigma_{\mathrm{E}}^t \\ \mu_{\mathrm{st}}^t = \alpha \mu_{\mathrm{SC}} S_{\mathrm{PV}} \mu_{\mathrm{E}}^t \\ \sigma_{\mathrm{st}}^t = \alpha \mu_{\mathrm{SC}} S_{\mathrm{PV}} \sigma_{\mathrm{E}}^t \\ \mu_{\mathrm{sp}}^t = (1-\alpha)\eta_{\mathrm{sp}} \mu_{\mathrm{SC}} S_{\mathrm{PV}} \mu_{\mathrm{E}}^t \\ \sigma_{\mathrm{sp}}^t = (1-\alpha)\eta_{\mathrm{sp}} \mu_{\mathrm{SC}} S_{\mathrm{PV}} \sigma_{\mathrm{E}}^t \end{cases} \tag{6.20}$$

The probability density functions (PDFs) for photovoltaic generation, solar thermal generation, and power generation can be determined:

$$f_{\mathrm{PV}}\left(P_{\mathrm{PV}}^{t}\right)=\frac{1}{\sigma_{\mathrm{PV}}^{t}\sqrt{2\pi}}e^{-\frac{\left(P_{\mathrm{PV}}^{t}-\mu_{\mathrm{PV}}^{t}\right)^{2}}{2\left(\sigma_{\mathrm{PV}}^{t}\right)^{2}}} \tag{6.21}$$

$$f_{\mathrm{st}}\left(H_{\mathrm{st}}^{t}\right)=\frac{1}{\sigma_{\mathrm{st}}^{t}\sqrt{2\pi}}e^{-\frac{\left(H_{\mathrm{st}}^{t}-\mu_{\mathrm{st}}^{t}\right)^{2}}{2\left(\sigma_{\mathrm{st}}^{t}\right)^{2}}} \tag{6.22}$$

$$f_{\mathrm{sp}}\left(P_{\mathrm{sp}}^{t}\right)=\frac{1}{\sigma_{\mathrm{sp}}^{t}\sqrt{2\pi}}e^{-\frac{\left(P_{\mathrm{sp}}^{t}-\mu_{\mathrm{sp}}^{t}\right)^{2}}{2\left(\sigma_{\mathrm{sp}}^{t}\right)^{2}}} \tag{6.23}$$

Setting the confidence levels to 95%, i.e., the given value $\omega = 0.05$ and $Z_{1-\omega/2} = 1.96$ (Z is the statistical data of photovoltaic, solar thermal and power generation), and the confidence intervals are:

$$\frac{P_{\mathrm{PV}}^{t}-\mu_{\mathrm{PV}}^{t}}{\sigma_{\mathrm{PV}}^{t}}\in\left[-Z_{1-\omega/2},Z_{1-\omega/2}\right]\Rightarrow P_{\mathrm{PV}}^{t}\in\left[\mu_{\mathrm{PV}}^{t}-1.96\sigma_{\mathrm{PV}}^{t},\mu_{\mathrm{PV}}^{t}+1.96\sigma_{\mathrm{PV}}^{t}\right] \tag{6.24}$$

$$\frac{H_{\mathrm{st}}^{t}-\mu_{\mathrm{st}}^{t}}{\sigma_{\mathrm{st}}^{t}}\in\left[-Z_{1-\omega/2},Z_{1-\omega/2}\right]\Rightarrow H_{\mathrm{st}}^{t}\in\left[\mu_{\mathrm{st}}^{t}-1.96\sigma_{\mathrm{st}}^{t},\mu_{\mathrm{st}}^{t}+1.96\sigma_{\mathrm{st}}^{t}\right] \tag{6.25}$$

$$\frac{P_{\mathrm{sp}}^{t}-\mu_{\mathrm{sp}}^{t}}{\sigma_{\mathrm{sp}}^{t}}\in\left[-Z_{1-\omega/2},Z_{1-\omega/2}\right]\Rightarrow P_{\mathrm{sp}}^{t}\in\left[\mu_{\mathrm{sp}}^{t}-1.96\sigma_{\mathrm{sp}}^{t},\mu_{\mathrm{sp}}^{t}+1.96\sigma_{\mathrm{sp}}^{t}\right] \tag{6.26}$$

Subsequently, the EPNS for the uncertain solar thermal, photovoltaic, and power generation at time t are:

$$P_{\mathrm{EPNS,PV}}^{t}=\int_{\mu_{\mathrm{PV}}^{t}-1.96\sigma_{\mathrm{PV}}^{t}}^{P_{\mathrm{PV}}^{t}}\left(P_{\mathrm{PV}}^{t}-x\right)f_{\mathrm{PV}}(x)dx \tag{6.27}$$

$$P_{\mathrm{EPNS,st}}^{t}=\int_{\mu_{\mathrm{st}}^{t}-1.96\sigma_{\mathrm{st}}^{t}}^{H_{\mathrm{st}}^{t}}\left(H_{\mathrm{st}}^{t}-x\right)f_{\mathrm{st}}(x)dx \tag{6.28}$$

$$P_{\mathrm{EPNS,sp}}^{t}=\int_{\mu_{\mathrm{sp}}^{t}-1.96\sigma_{\mathrm{sp}}^{t}}^{P_{\mathrm{sp}}^{t}}\left(P_{\mathrm{sp}}^{t}-x\right)f_{\mathrm{sp}}(x)dx \tag{6.29}$$

The EPC for the uncertain solar thermal, photovoltaic, and power generation at time t are given by the following expressions:

$$P^t_{\mathrm{EPC,PV}} = \int\limits_{P^t_{\mathrm{PV}}}^{\mu^t_{\mathrm{PV}}+1.96\sigma^t_{\mathrm{PV}}} \left(x - P^t_{\mathrm{PV}}\right) f_{\mathrm{PV}}(x)dx \tag{6.30}$$

$$P^t_{\mathrm{EPC,st}} = \int\limits_{H^t_{\mathrm{st}}}^{\mu^t_{\mathrm{st}}+1.96\sigma^t_{\mathrm{st}}} \left(x - H^t_{\mathrm{st}}\right) f_{\mathrm{st}}(x)dx \tag{6.31}$$

$$P^t_{\mathrm{EPC,sp}} = \int\limits_{P^t_{\mathrm{sp}}}^{\mu^t_{\mathrm{sp}}+1.96\sigma^t_{\mathrm{sp}}} \left(x - P^t_{\mathrm{sp}}\right) f_{\mathrm{sp}}(x)dx \tag{6.32}$$

In the third step, we address the high risk posed by irradiation intensity uncertainty on the operation of the agricultural microgrid by incorporating the uncertainty costs associated with the photovoltaic and solar thermal systems. These costs, including EPNS and EPC costs, are integrated into the overall cost structure of the agricultural microgrid and are expressed as follows:

$$\begin{aligned} C_{\mathrm{PV}} = &\sum_{t=1}^{T}\left(\lambda^t P^t_{\mathrm{EPNS,PV}} + \gamma^t P^t_{\mathrm{EPNS,st}} + \lambda^t P^t_{\mathrm{EPNS,sp}}\right) \\ &+ \sum_{t=1}^{T}\lambda_0\left(P^t_{\mathrm{EPC,PV}} + P^t_{\mathrm{EPC,st}} + P^t_{\mathrm{EPC,sp}}\right) \end{aligned} \tag{6.33}$$

where λ^t and γ^t are the electricity and natural gas price, respectively; λ_0 is the cost coefficient for photovoltaic abandonment.

The risk costs associated with EPNS and EPC, as described in Eq. (6.33), provide a mechanism to balance photovoltaic and solar thermal output and abandonment. Specifically, EPNS quantifies the anticipated shortfall in energy supply to users, helping to mitigate the risk of insufficient energy supply. In contrast, EPC measures the energy loss caused by photovoltaic abandonment, aiming to enhance the economic efficiency of the agricultural microgrid's operation.

6.3.2 *Resilience of Agricultural Microgrid*

Ensuring continuous provision of water, electricity, and heat under low-probability, high-impact events is crucial for the quality of life and economic development. Securing load supply is fundamental for the stable operation of agricultural microgrids. Therefore, a key objective of the EHWbMEH model is to leverage various resources to enhance the resilience of the agricultural microgrid, ensuring that loads are as secure as possible. In this chapter, the resilience indices R_1, R_2 and

R_3 are defined according to Eqs. (6.34), (6.35), and (6.36). Maximizing these indices improves the resilience of the agricultural microgrid during supply outages.

$$R_1 = \int_0^T \frac{P_{\mathrm{r}}^t}{P_{\mathrm{load}}^t} dt \tag{6.34}$$

$$R_2 = \int_0^T \frac{H_{\mathrm{r}}^t}{H_{\mathrm{load}}^t} dt \tag{6.35}$$

$$R_3 = \int_0^T \frac{Q_{\mathrm{r}}^t}{Q_{\mathrm{Ir}}^t} dt \tag{6.36}$$

where P_{r}^t, H_{r}^t and Q_{r}^t in the EHWbMEH model quantify the capacity of the system to maintain load supply during interruptions in electricity, gas, and/or water supplies, respectively.

In addition, to ensure maximum load supply, we introduce a penalty cost C_{pen}, calculated as follows:

$$C_{\mathrm{pen}} = \sum_{t=1}^{T} \left(P_{\mathrm{load}}^t - P_{\mathrm{r}}^t\right)\eta_{\mathrm{P}} + \sum_{t=1}^{T} \left(H_{\mathrm{load}}^t - H_{\mathrm{r}}^t\right)\eta_{\mathrm{H}} + \left(Q_{\mathrm{TW}} - \sum_{t=1}^{T} Q_{\mathrm{r}}^t\right)\eta_{\mathrm{Q}} \tag{6.37}$$

where η_{P}, η_{H} and η_{Q} denote the penalty coefficients for losing of electricity, heat and water loads, respectively.

6.3.3 Operation Cost of Agricultural Microgrid

While ensuring that the demands for electricity, heat, and water loads of the agricultural microgrid are met, the objective is to minimize the total operational cost by optimizing the performance. The objective for calculating the total operational cost of the agricultural microgrid is:

$$C = \sum_{t=1}^{T} \left(\lambda^t P_{\mathrm{N}}^t + \gamma^t Q_{\mathrm{gas}}^t\right) + C_{\mathrm{PV}} + C_{\mathrm{pen}} \tag{6.38}$$

The total operational cost, as outlined in Eq. (6.38), is comprised of three components: the first component represents the purchasing electricity and natural gas cost; the second component accounts for the uncertainty costs associated with photovoltaic and solar thermal systems; and the final component reflects the penalty cost for losing of loads.

Let P_{r}^{t}, H_{r}^{t} and Q_{r}^{t} denote the critical loads for electricity, gas, and water demands, respectively. This chapter addresses minimization of operational costs and maximization of resilience. The optimization model is formulated as follows:

$$\min \quad [C, -\min\{R_1, R_2, R_3\}]\text{s.t.}\begin{cases} \underbrace{(1)-(19)}_{\text{EHWbMEH constraints}} \\ \underbrace{(24)-(26)}_{\text{Confidence intervals}} \\ P_{\mathrm{r}}^{t} \geq P_{\mathrm{c}}^{t}, H_{\mathrm{r}}^{t} \geq H_{\mathrm{c}}^{t}, Q_{\mathrm{r}}^{t} \geq Q_{\mathrm{c}}^{t} \end{cases} \tag{6.39}$$

6.3.4 Multiple Preys Based Evolutionary Predator and Prey Strategy

MPEPPS is an evolutionary algorithm inspired by the survival mechanisms of animals in nature. It is well-suited for solving multi-objective optimization problems, offering effective Pareto solutions by balancing local and global search capabilities.

In this chapter, the following steps outline the procedure for applying the MPEPPS to the agricultural microgrid optimization problem: Initially, input the data, such as predicted irradiation intensity and ambient temperature. Subsequently, the parameters of the MPEPPS are configured. During each iteration, members of the algorithm perform distinct roles and utilize different searching mechanisms. The Pareto solutions are updated using a fast non-dominated sorting method, which is based on the fitness values calculated from the location vectors of each member. The algorithm iterates until a predefined condition is met, at which point the computation concludes. The computational procedures of MPEPPS are illustrated in Fig. 6.2.

6.4 Case Study

6.4.1 Parameter Settings

The panel area for the photovoltaic and solar thermal systems in the agricultural microgrid is set at 300 m^2, with the cell temperature stabilized at 35 °C. The predicted irradiation intensity, as well as heat and electricity demands, are illustrated in Fig. 6.3 [29]. Water demand data and hydro-electric conversion coefficients for the pumped storage system and irrigation system are sourced from [30]. The CHP unit has conversion efficiencies of 0.35 for electricity and 0.65 for heat, while the HP unit has an electricity conversion coefficient of 3. Operational constraints for each conversion device are listed in Table 6.1. Critical loads are set at 10% of the total demands.

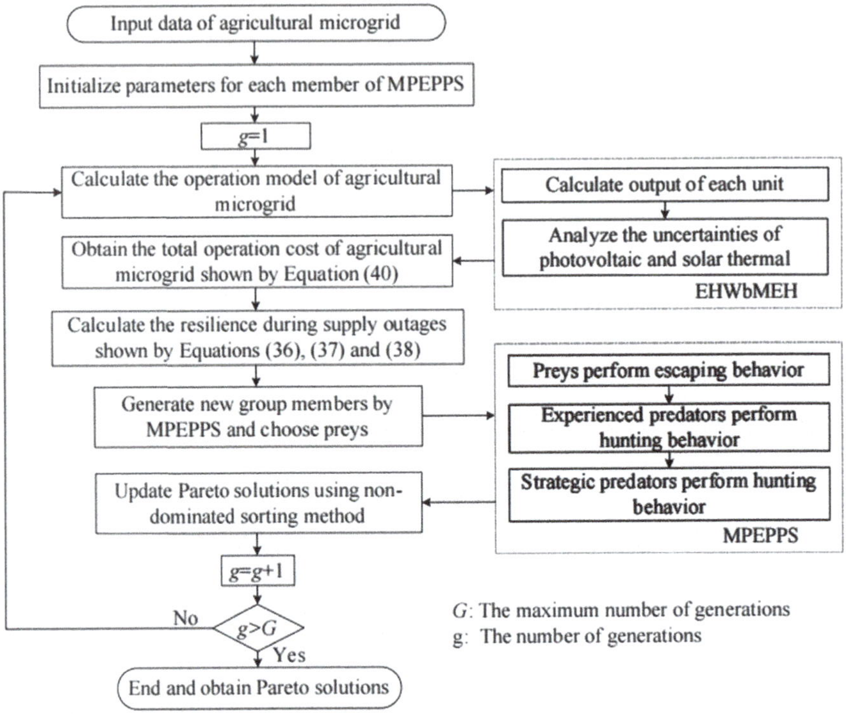

Fig. 6.2 The computing procedures of MPEPPS, reprinted from ref. [26]. Copyright 2024, with permission Renewable Energy

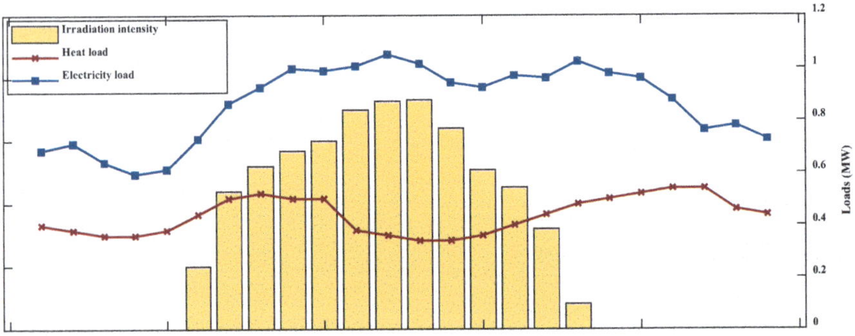

Fig. 6.3 The predicted irradiation intensity and load demands of agricultural microgrid

Electricity prices from the upstream grid follow a sharp-peak-flat-valley pricing scheme: \$190/MWh from 15:00 to 21:00 (sharp peak), \$150/MWh from 9:00 to 14:00 (peak), \$135/MWh from 7:00 to 8:00 and 22:00 to 24:00 (flat), and \$114/MWh from 1:00 to 6:00 (valley) [30]. Natural gas prices are based on a time-of-use rate: \$50/MWh from 1:00 to 6:00, \$80/MWh from 9:00 to 20:00, and \$60/MWh

Table 6.1 Variable bounds of the proposed EHWbMEH

Variable	Interval (Hm^3)	Variable	Interval (MW)
V_{up}	[0.05, 0.5]	P_{HP}	[0, 0.5]
V_{dn}	[0.1, 0.7]	Q_{gas}	[0, 0.5]
Q_P	[0.01, 0.06]	H	[0, 1.5]
Q_G	[0.01, 0.06]	H_{ch}	[− 0.5, 0.5]
Q_{Ir}	[0.005, 0.02]		

from 7:00 to 8:00 and 21:00 to 24:00 [31]. Additionally, penalty costs for unsupplied electricity, heat, and water loads are set at \$600/MWh, \$550/MWh, and \$500/Hm^3, respectively.

6.4.2 *Influence of EHWbMEH on Agricultural Microgrid Operation*

Two scenarios are presented to verify the proposed EHWbMEH. **Case 1**: The agricultural microgrid is optimized with the EHWbMEH in operation, ensuring that each component meets the electricity, heat, and water demands. This case considers the interdependencies between the power system, thermal system, and irrigation system, reflecting their combined operation in supplying the necessary loads. **Case 2**: The systems operate independently without the EHWbMEH. In this scenario, all heat collected by the solar thermal system is directed to the thermal system, while the power needed for the irrigation system and the HP unit is directly purchased from the upstream grid. This case isolates the performance of each system, lacking the integrated synergy of the EHWbMEH.

Figure 6.4 illustrates the operational parameters of the pumped storage system, HP unit, irrigation system, and TSU in both Case 1 and Case 2. In Case 1, during hours 1–6, when electricity prices are low, the pumped storage system transfers water from the lower reservoir to the upper reservoir, thereby storing surplus electricity. Conversely, during hours 7–24, when electricity prices are high, the system releases water to generate electricity and meet the electricity demand. In Case 2, the pumped storage system's operation is limited to supplying water for the irrigation system. The volume of released water remains constant, as there is no involvement in energy storage or generation.

In Case 1, the electricity consumption of the HP rises notably during hours 9, 16, 18, and 20, when heat demand is high. Conversely, during hours 1–6 and 22–24, when heat demand is low, the CHP unit adequately meets the heat requirements, resulting in reduced HP output.

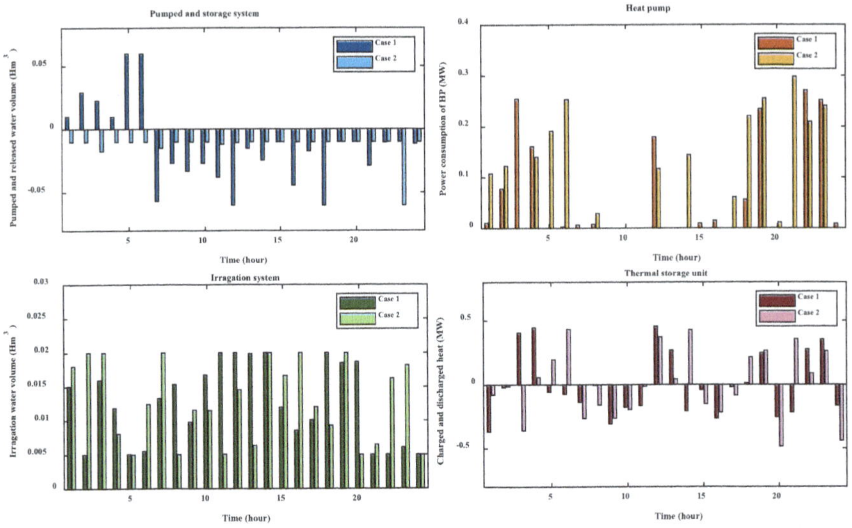

Fig. 6.4 Equipment operations in cases 1 and 2

In Case 2, the separation of the power and thermal systems and the non-operation of the CHP unit lead to a substantial increase in HP output. The total power consumption in this case is 2.968 MW, representing a 65.5% increase compared to Case 1.

TSU enhances peak-shaving and valley-filling, thereby balancing heat consumption and production and providing greater economic benefits to the agricultural microgrid. During hours 6, 9, and 12, the HP and CHP units produce more heat than is required, causing the TSU to absorb the excess heat. Notably, in hour 9, the heat stored in the TSU reaches 0.492 MW. In hours 10–15, when the HP and CHP output is insufficient to meet heat demand, the TSU supplies additional heat to the microgrid.

Furthermore, water consumption is inversely related to power prices in irrigation system. The volume of water used by the irrigation system decreases significantly during hours 15–20, when power prices are higher.

Table 6.2 evaluates the effect of EHWbMEH on the power purchased from the upstream grid and the overall operation cos. The analysis reveals that the total operation cost in Case 1 is $3,170.134, which is $431.530 lower than in Case 2. In Case 1, the power purchased from the upstream grid is reduced by 6.067 MW compared to Case 2. This reduction is attributed to the effective utilization of the CHP unit, photovoltaic, and solar thermal systems, which leverage the multi-energy complementarity within the EHWbMEH framework to minimize the total operation cost. Consequently, EHWbMEH enables the integrated management of electricity, heat, and water resources, significantly lowering the operational expenses of the agricultural microgrid.

Table 6.2 Optimization results in cases 1 and 2

Hour	Case power purchase (MW)	1 operation cost ($)	Case power purchase (MW)	2 operation cost ($)
1	0.820	93.481	0.684	100.462
2	0.889	111.459	0.711	106.423
3	1.087	123.941	0.639	84.244
4	0.688	103.450	0.594	88.203
5	0.762	111.892	0.612	94.432
6	0.867	121.679	0.716	117.616
7	0.363	79.115	0.830	125.640
8	0.649	114.991	0.874	125.433
9	0.670	119.485	0.953	151.699
10	0.680	129.433	0.930	148.120
11	0.674	124.049	0.932	143.715
12	0.952	142.946	0.967	173.593
13	0.792	151.201	0.921	142.965
14	0.743	122.718	0.860	165.674
15	0.828	174.738	0.858	183.613
16	0.737	143.875	0.910	196.961
17	0.701	176.839	0.917	202.429
18	0.698	175.866	1.017	250.493
19	1.233	246.115	0.981	260.595
20	0.862	200.875	0.963	194.901
21	0.588	144.680	0.882	236.475
22	1.007	136.278	0.765	142.537
23	0.985	138.180	0.783	150.460
24	0.569	100.056	0.729	101.791
Total	18.844	3287.342	20.028	3688.474

To further validate the proposed optimization, numerical experiments conducted on a modified IEEE 33-node power distribution network (PDN), a 32-node district heating network (DHN), and a 15-node water distribution network (WDN) have been studied. Detailed system data and topological information are provided in [29, 32]. To ensure the robustness and reliability of our results, each test was independently repeated 30 times. The results of these tests are summarized as follows.

For a scenario with electricity interruption occurs between hours 12 and 13, Fig. 6.5 illustrates the outputs of CHP and pumped storage system within EHWbMEH, as well as the storage volumes. The figure demonstrates that the pumped storage system efficiently releases water during peak load periods and pumps it during off-peak hours. Specifically, during the outage, when the released power generation peaks, the CHP also operates at its maximum capacity. Under this scenario, the total

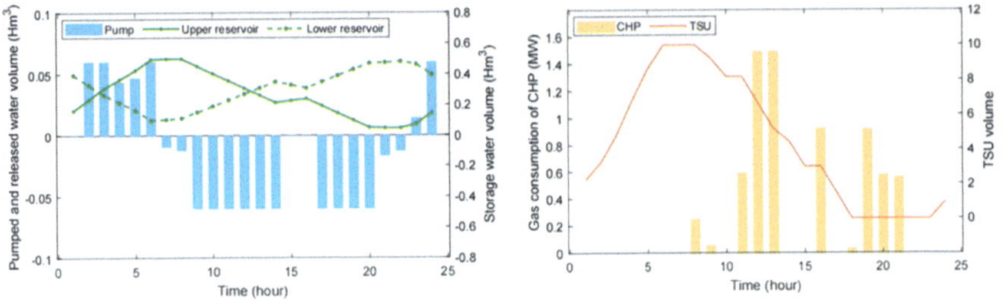

Fig. 6.5 Equipment operations of the optimal Pareto solution

operation cost is $8,075.98, and the electricity load resilience is 0.4356. In response to the electricity outage, EHWbMEH prioritizes the supply of electricity to critical loads, with the CHP effectively converting natural gas into electricity and heat. This approach minimizes heat load losses despite power supply constraints, thereby enhancing the system's overall resilience during emergencies.

6.4.3 Analysis of Agricultural Microgrid Resilience Enhancement

Figure 6.6 presents the illustrating interruptions in electricity, gas, and water supply under extreme scenarios for agricultural microgrids. The chapter examines resilience enhancement under the following scenarios:

Scenario 1: Electricity interruption√, Water interruption×, Gas interruption×
Scenario 2: Electricity interruption×, Water interruption×, Gas interruption√
Scenario 3: Electricity interruption×, Water interruption√, Gas interruption×
Scenario 4: Electricity interruption√, Water interruption×, Gas interruption√
Scenario 5: Electricity interruption√, Water interruption√, Gas interruption×

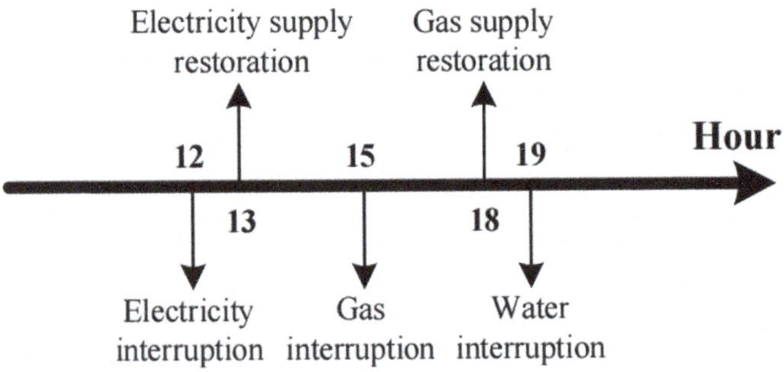

Fig. 6.6 Time series events diagram, reprinted from ref. [26]. Copyright 2024, with permission renewable energy

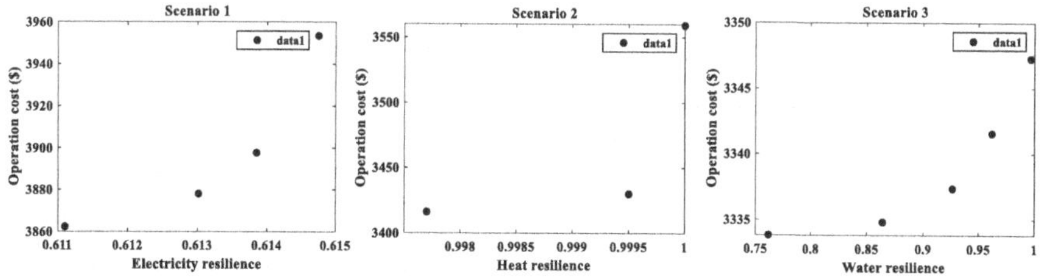

Fig. 6.7 Pareto frontiers for Scenarios 1, 2 and 3

Scenario 6: Electricity interruption×, Water interruption√, Gas interruption√

Scenario 7: Electricity interruption√, Water interruption√, Gas interruption√

Figure 6.7 displays the Pareto frontiers for Scenarios 1, 2, and 3. When electricity supply is interrupted, the system can maintain less electricity load, with 6% of the electricity load remaining operational due to the effects of the pumped storage system and CHP unit. However, in cases where only natural gas or water supply interruptions occur, the resilience for heat and water loads can reach 100%.

Figure 6.8 illustrates the Pareto frontiers and corresponding resilience for Scenarios 4, 5, and 6. As the severity of the events escalates, maintaining the supply for the agricultural microgrid becomes increasingly difficult, leading to a rise in the total operational cost. Notably, the penalty costs for unsupplied loads are substantially high, amounting to $447.627 for Scenario 4 and $417.940 for Scenario 5.

Figure 6.9 presents the corresponding Pareto frontier and resilience metrics of Scenario 7. The optimal trade-off solution in this scenario results in a total operational cost of $3740.141. The resilience levels are 63.37% for electricity, 84.72% for heat, and 80.55% for water. During the electricity outage, CHP achieves peak outputs of 0.499 MW and 0.492 MW at the 12th and 13th hours, respectively. Additionally, the solar collector effectively converts most of the collected heat into electricity to sustain the electricity load as much as possible.

Figure 6.10 shows the storage volumes for the TSU and pumped storage system. During the electricity interruptions, the water flow from the pumped storage system increases significantly to offset the lack of electricity supply. Concurrently, the TSU releases more heat to reduce the power consumption of HP. These results illustrate that EHWbMEH approach can significantly enhance the resilience during supply disruptions while also decreasing the total operational cost.

6.4.4 Utility Analysis of Photovoltaic and Solar Thermal System

Figure 6.11 depicts the Pareto frontier and resilience metrics of an agricultural microgrid with an uncoupled photovoltaic and solar thermal system. The optimal trade-off solution results in a total operational cost of $3767.400, with resilience levels of

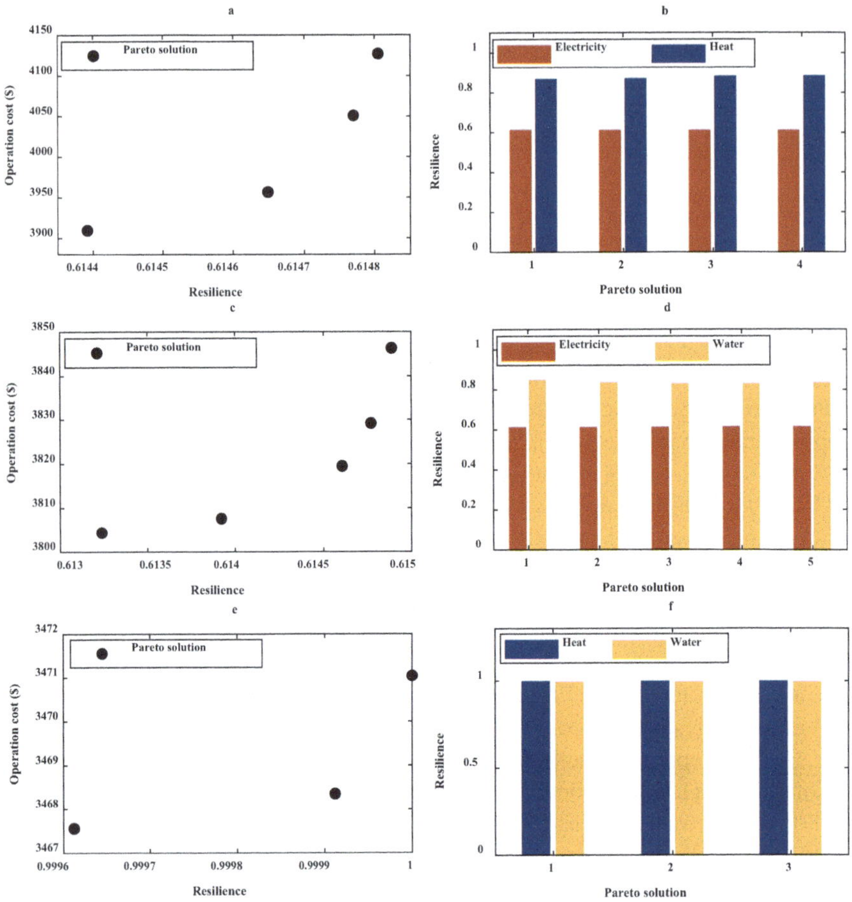

Fig. 6.8 Pareto frontiers and the corresponding resilience for Scenarios 4, 5, and 6

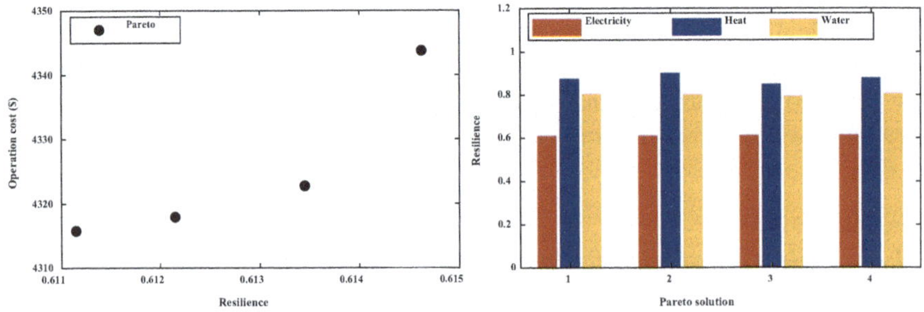

Fig. 6.9 Pareto frontier and the corresponding resilience for Scenario 7

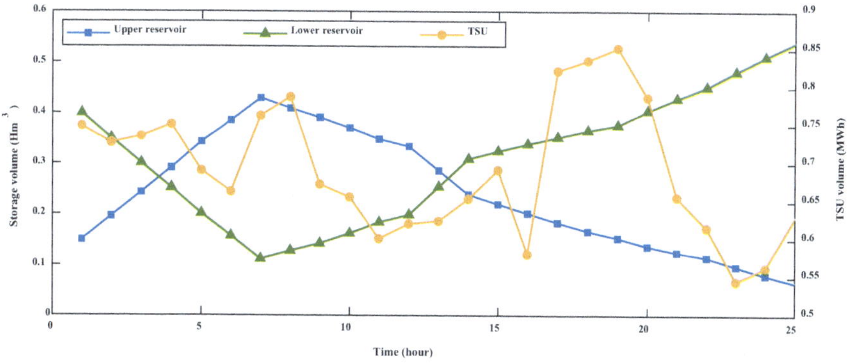

Fig. 6.10 The storage volume of pumped storage system and TSU

53.37% for electricity, 81.70% for heat, and 77.73% for water. However, compared to the coupled system, the resilience for electricity, heat, and water decreases. The electricity resilience decreases by approximately 10%. This finding indicates that coupling photovoltaic and solar thermal systems substantially enhances agricultural microgrid resilience.

Table 6.3 provides optimal outputs with coupled and uncoupled photovoltaic and solar thermal systems. The results show that utilizing the coupled system enhances the resilience of the microgrid in terms of electricity, heat, and water supply while reducing cost by $27.259. Additionally, the power generation of the photovoltaic-thermal system increases by 0.644 MW, allowing the microgrid to supply more electricity load during interruptions.

To investigate the effects of varying errors, the photovoltaic and solar thermal output coefficients, and the abandonment coefficient are set to $600/MW. Figure 6.12 illustrates the Pareto solutions for operation cost and resilience across errors ranging from 1 to 5%.

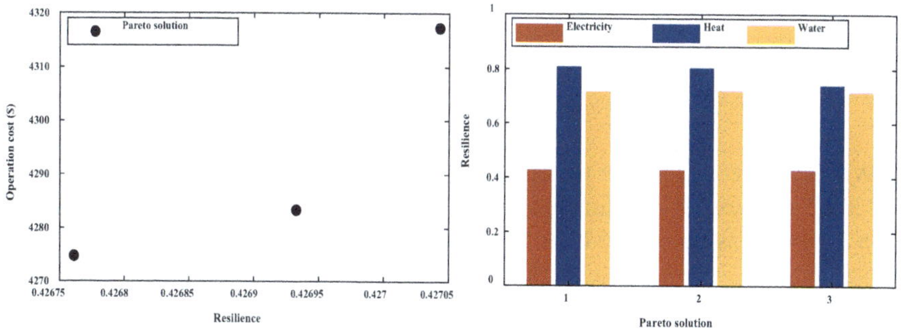

Fig. 6.11 Pareto frontier and the corresponding resilience for uncoupled photovoltaic and solar thermal

Table 6.3 Comparison of optimization results with coupled and uncoupled photovoltaic and solar thermal system

Hour	Coupled system			Uncoupled system		
	Operation cost ($)	Power generation (MW)	Heat generation (MW)	Operation cost ($)	Power generation (MW)	Heat generation (MW)
1	124.849	0	0	125.255	0	0
2	129.036	0	0	124.846	0	0
3	119.528	0	0	123.345	0	0
4	114.941	0	0	119.945	0	0
5	111.937	0	0	113.664	0	0
6	130.080	0.035	0.058	130.448	0.020	0.095
7	117.220	0.090	0.050	137.515	0.048	0.212
8	123.053	0.117	0.026	139.860	0.055	0.244
9	144.386	0.101	0.122	174.740	0.059	0.271
10	140.002	0.118	0.083	184.937	0.065	0.272
11	138.785	0.168	0.032	168.641	0.072	0.330
12	295.562	0.197	0.038	415.142	0.095	0.347
13	276.469	0.197	0.037	393.364	0.095	0.327
14	131.152	0.131	0.093	159.748	0.069	0.309
15	174.271	0.100	0.082	198.510	0.056	0.236
16	214.315	0.095	0.067	205.214	0.048	0.207
17	198.133	0.060	0.072	242.007	0.031	0.143
18	229.699	0.012	0.021	250.928	0.008	0.035
19	192.100	0	0	195.175	0	0
20	191.978	0	0	189.993	0	0
21	172.938	0	0	171.139	0	0
22	109.787	0	0	108.398	0	0
23	109.983	0	0	111.642	0	0
24	106.607	0	0	101.489	0	0
Total	3796.811	3796.811	3796.811	3796.811	3796.811	3796.811

With an error of 1%, the resilience stands at 0.634, while the operation cost is \$4320.62. As the error escalates to 5%, the resilience improves to 0.667, accompanied by a cost increasing of \$4540.49. This highlights the pivotal role played by the EPNS and EPC values in balancing the photovoltaic and solar thermal output and abandonment within the agricultural microgrid. While an increase in the error leads to an enhancement in resilience, it also results in a rise in the operational cost.

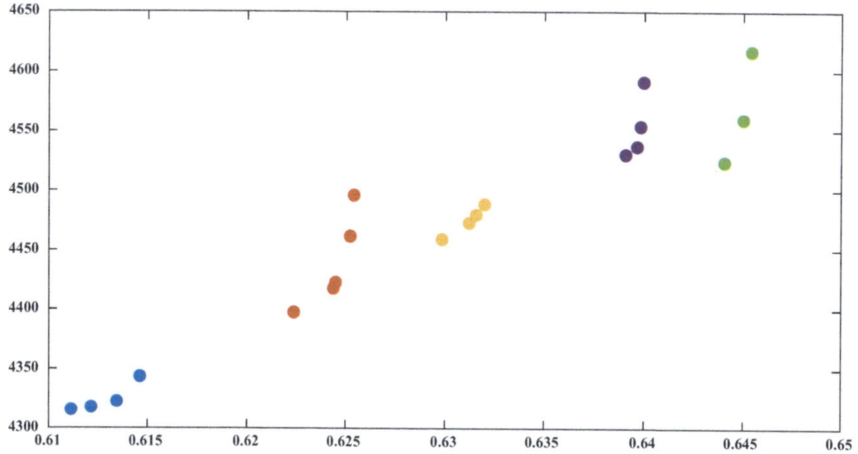

Fig. 6.12 Pareto frontier for different prediction errors

6.5 Conclusion

This chapter presents an EHWbMEH for resilience-economic agricultural microgrids optimization, incorporating the uncertainty of irradiation intensity. The efficiency of the proposed optimization are demonstrated through comprehensively studies.

(1) The EHWbMEH framework achieves operation cost reduction by $431.530. The TSU and pumped storage system act as energy storage, facilitating peak-shaving and valley-filling, and enhancing the efficiency of photovoltaic and solar thermal utilization.
(2) The integrated use of photovoltaic and solar thermal technologies in agricultural microgrids fully leverages the synergistic potential of power and thermal systems, maximizing the utilization of solar resources and enhancing the resilience of agricultural microgrids during supply outages.
(3) The resilience-economic optimization balances enhancing resilience and reducing cost of agricultural microgrids. In scenarios with interrupted electricity, heat, and water supply, the trade-off solution results in a cost of $3740.141, with electricity resilience at 63.37%, heat resilience at 84.72%, and water resilience at 80.55%.

References

1. Kama MM, Ashraf I, Fernandez E (2022) Optimal planning of renewable integrated rural microgrid for sustainable energy supply. Energy Storage 4(4):e332
2. Tan H, Li Z, Wang Q, Mohamed MA (2023) A novel forecast scenario-based robust energy management method for integrated rural energy systems with greenhouses. Appl Energy 330:120343
3. Wang Y, Dong W, Yang Q (2022) Multi-stage optimal energy management of multi-energy microgrid in deregulated electricity markets. Appl Energy 310:118528
4. Zhong X, Zhong W, Liu Y, Yang C, Xie S (2022) Optimal energy management for multi-energy multi-microgrid networks considering carbon emission limitations. Energy 246:123428
5. Shao Z, Cao X, Zhai Q, Guan X (2023) Risk-constrained planning of rural-area hydrogen-based microgrid considering multiscale and multi-energy storage systems. Appl Energy 334:120682
6. Honarmand HA, Rashid SM (2022) A sustainable framework for long-term planning of the smart energy hub in the presence of renewable energy sources, energy storage systems and demand response program. J Energy Storage 52:105009
7. Akbari E, Shabestari SFM, Pirouzi S, Jadidoleslam M (2023) Network flexibility regulation by renewable energy hubs using flexibility pricing-based energy management. Renew Energy 206:295–308
8. Gharai MA, Hooshmand RA, Nosratabadi SM et al (2023) Eco-environmental uncertain management of various energy hubs within a microgrid concentrated on demand response, intelligent parking lots, and power-to-gas unit effects. Int J Energy Res 2023
9. Mokaramian E, Shayeghi H, Sedaghati F, Safari A (2021) Four-objective optimal scheduling of energy hubusing a novel energy storage, considering reliability and risk indices. J Energy Storage 40:102731
10. Wang L, Dong H, Lin J, Zeng M (2022) Multi-objective optimal scheduling model with IGDT method of integrated energy system considering ladder-type carbon trading mechanism. Int J Electr Power Energy Syst 143:108386
11. Meng Z, Dong F, Chi L (2023) Optimal dispatch of integrated energy system with p2g considering carbon trading and demand response. Environ Sci Pollut Res 1–20
12. Hashemi S, Arasteh H, Shafiekhani M, Kia M, Guerrero J (2023) Multi-objective operation of microgrids based on electrical and thermal flexibility metrics using the NNC and IGDT methods. Int J Electr Power Energy Syst 144:108617
13. Mousavi N, Kothapalli G, Habibi D, Das CK, Baniasadi A (2020) A novel photovoltaic-pumped hydrostorage microgrid applicable to rural areas. Appl Energy 262:114284
14. Amirioun MH, Aminifar F, Shahidehpour M (2018) Resilience-promoting proactive scheduling against hurricanes in multiple energy carrier microgrids. IEEE Trans Power Syst 34(3):2160–2168
15. Huang W, Du E, Capuder T, Zhang X, Zhang N, Strbac G et al (2021) Reliability and vulnerability assessment of multi-energy systems: an energy hub based method. IEEE Trans Power Syst 36(5):3948–3959
16. Mohseni S, Brent AC (2022) Quantifying the effects of forecast uncertainty on the role of different battery technologies in grid-connected solar photovoltaic/wind/micro-hydro microgrids: an optimal planning study. J Energy Storage 51:104412
17. Liu J, Chen J, Yan G, Chen W, Xu B (2023) Clustering and dynamic recognition based autoreservoir neural network: a wait-and-see approach for short-term park power load forecasting. Iscience 26(8)
18. Wang W, Yang D, Huang N, Lyu C, Zhang G, Han X (2022) Irradiance-to-power conversion based on physical model chain: an application on the optimal configuration of multi-energy microgrid in cold climate. Renew Sustain Energy Rev 161:112356
19. Yang J, Su C (2021) Robust optimization of microgrid based on renewable distributed power generation and load demand uncertainty. Energy 223:120043

20. Dong W, Yang Q, Fang X, Ruan W (2021) Adaptive optimal fuzzy logic based energy management in multi-energy microgrid considering operational uncertainties. Appl Soft Comput 98:106882
21. Luo L, Abdulkareem SS, Rezvani A, Miveh MR, Samad S, Aljojo N et al (2020) Optimal scheduling of a renewable based microgrid considering photovoltaic system and battery energy storage under uncertainty. J Energy Storage 28:101306
22. Ali A, Liu N, He L (2019) Multi-party energy management and economics of integrated energy microgrid with PV/T and combined heat and power system. IET Renew Power Gener 13(3):451–461
23. Zhang X, Wang Z, Lu Z (2022) Multi-objective load dispatch for microgrid with electric vehicles using modified gravitational search and particle swarm optimization algorithm. Appl Energy 306:118018
24. Shahryari E, Shayeghi H, Mohammadi-Ivatloo B, Moradzadeh M (2019) A copula-based method to consider uncertainties for multi-objective energy management of microgrid in presence of demand response. Energy 175:879–890
25. Ramli MA, Bouchekara H, Alghamdi AS (2018) Optimal sizing of PV/wind/diesel hybrid microgrid system using multi-objective self-adaptive differential evolution algorithm. Renew Energy 121:400
26. Zhang YQ, Chen JJ, Wang YX et al (2024) Enhancing resilience of agricultural microgrid through electricity–heat–water based multi-energy hub considering irradiation intensity uncertainty[J]. Renew Energy 220:119739
27. Lu Y, Li G (2023) Potential application of electrical performance enhancement methods in PV/T module. Energy 281:128253
28. Sun K, Li KJ, Pan J, Liu Y, Liu Y (2019) An optimal combined operation scheme for pumped storage and hybrid wind-photovoltaic complementary power generation system. Appl Energy 242:1155
29. Li R, Wei W, Mei S, Hu Q, Wu Q (2018) Participation of an energy hub in electricity and heat distribution markets: an MPEC approach. IEEE Trans Smart Grid 10(4):3641–3653
30. Zhang M, Chen J, Yang Z, Peng K, Zhao Y, Zhang X (2021) Stochastic day-ahead scheduling of irrigation system integrated agricultural microgrid with pumped storage and uncertain wind power. Energy 237:121638
31. Zhao B, Conejo AJ, Sioshansi R (2016) Unit commitment under gas-supply uncertainty and gas-price variability. IEEE Trans Power Syst 32(3):2394–2405
32. Rodriguez-Garcia L, Hosseini MM, Mosier TM, Parvania M (2022) Resilience analytics for interdependent power and water distribution systems. IEEE Trans Power Syst 37(6):4244–4257

Chapter 7
Continuous-Time Optimization to Improve Demand Defense of Renewable Industrial Park

7.1 Introduction

Under the drive of the "dual-carbon" goals put forward by the Chinese government, photovoltaic (PV) power generation and energy storage are leading to a revolutionary change in electric power technologies [1–3]. Currently, the cost of PV power generation has fallen below that of coal-fired electricity prices. Industrial and commercial parks can take advantage of their facilities, rooftops, carports, and other resources to install PV systems, achieving self-consumption while feeding excess electricity back to the grid [4]. This not only reduces electricity expenses but also decreases carbon emissions, alleviating power supply shortages. Furthermore, the application of energy storage systems in parks can further reduce electricity expenses, improve the proportion of on-site consumption of PV power, enhance the utilization efficiency of grid equipment, and ensure uninterrupted power supply for critical loads during islanded operation.

In line with these developments, the National Development and Reform Commission (NDRC) and the National Energy Administration (NEA) issued Document No. (2021) 1051, encouraging the application of energy storage systems in industrial parks, large data centers, 5G base stations, microgrids, and other end-users [5]. Document No. (2021) 1093, also released by the NDRC, proposes to further improve peak-valley electricity pricing mechanisms, continuously widen the price difference between peak and off-peak electricity, and raise peak electricity prices. This aims to encourage industrial and commercial users to reduce electricity loads during peak periods and increase electricity consumption during off-peak periods by deploying energy storage systems.

Against this backdrop, user-side photovoltaic and energy storage (PV-ESS) parks are poised to become integral components of the new power system. They play a proactive role in attracting social capital for investments in new energy projects, optimizing energy structures, and facilitating the peak carbon emissions and carbon neutrality goals [6, 7]. Therefore, the key challenge lies in how to timely adjust

J. Chen and Y. Li, *Renewable Power System Optimization*, Smart Energy Systems,
https://doi.org/10.1007/978-981-97-8132-4_7

the charging and discharging power of energy storage systems during the dynamic changes of PV generation and loads, aiming to maximize the benefits of PV-ESS parks.

Currently, research on the operation and control of energy storage in PV-ESS parks has achieved some results, mainly focusing on peak-valley arbitrage, promoting on-site consumption of PV power, and reducing demand charges. For example, in the peak-valley arbitrage mode, Reference [8] determines the charging and discharging start and end times of energy storage based on time-of-use electricity prices and state-of-charge information. The charging and discharging power levels are then determined according to the actual load demand, achieving the economic operation of PV-ESS parks. Reference [9] shows that deploying energy storage on the user side with PV can increase the proportion of on-site consumption of PV power, reduce users' electricity expenses, and simultaneously increase investors' profits. Reference [10] proposes a method for assessing and optimizing park energy storage by bundling demand management with peak-valley arbitrage, providing an in-depth study of the economic aspects of demand defense.

Overall, in terms of peak-valley arbitrage, the operation mode of energy storage involves charging during off-peak hours when electricity prices are low and discharging during peak hours when prices are high [11, 12]. By balancing arbitrage benefits with operational costs, the optimal number of daily charge–discharge cycles for energy storage is determined. This operating mode is simple, with relatively low technical challenges. As for on-site consumption of PV power, considering the lower grid electricity prices for PV, when PV generation exceeds the park's load, the surplus PV power is prioritized for charging the energy storage system. This scenario is relatively straightforward, and the implementation is relatively easy.

Under the current widely adopted two-part tariff, the demand charge for industrial parks each month is equal to the average of the 15-min maximum power consumption values from the main meter multiplied by the demand charge rate. Generally, PV-ESS parks use prediction methods or historical data to obtain the future month's PV generation and load demand curves, and then optimize energy storage charging and discharging with the objective of minimizing operational costs [13]. This optimization determines the maximum monthly demand, known as the monthly demand power defense value.

However, in practical engineering scenarios, it is challenging to accurately predict the PV generation and load curves of the park, especially in extreme weather conditions, park line faults [11, 14], sudden startup of high-power experimental equipment, and other special scenarios. Taking Shandong province, China as an example, the demand power charge for 1–10 (20 kV) industrial and commercial users is 38.4 CNY/kW/month, with an additional cost of 1.28 CNY for each additional kW of demand, exceeding the Shandong province peak electricity price of 1.137 CNY/kW. Therefore, dynamically adjusting the energy storage charging and discharging power in response to the random changes in PV generation and load is a major challenge in the operation and control of energy storage in PV-ESS parks, particularly in effectively defending against the monthly demand power charges.

Although refining the operational methods at finer time scales can alleviate the pressure of demand charge defense for the park to some extent, such as 15-min or minute-level energy storage optimization, these methods have the following drawbacks [15, 16]. Firstly, they can only obtain the state of charge of the energy storage for a certain time segment, making it difficult to accurately capture the operational state of the energy storage within that time segment, resulting in inaccurate energy storage capacity requirements. Secondly, they cannot accurately reflect the ramping process of energy storage, which can easily cause a mismatch between supply and demand of energy for a certain time segment.

In recent years, continuous-time optimization models based on function approximation have garnered widespread attention in power system scheduling. Current research on continuous-time control strategies can embed optimization problems into fitted continuous functions. Continuous-time control strategies have been successfully applied in areas such as flexibility enhancement, energy storage configuration, and stochastic scheduling. Reference [17] proposed a flexibility enhancement model based on Bernstein polynomials aimed at exploring potential flexibility resources in power systems. Reference [18] presented a robust optimization model at the continuous-time scale to study the optimal configuration of energy storage devices in the case of high penetration of wind power generation. In Reference [10], the authors established a continuous-time stochastic optimization model for generators and energy storage to cooperatively optimize the electric energy and flexibility reserves provided by generators and energy storage devices. The optimization periods in the above studies are fixed, such as 1 h or 15 min. However, when there is a sudden increase in load or a fault occurs, the length of the optimization interval is often non-fixed. The continuous function fitting methods used in the above literature are difficult to capture the dynamic information in emergency scheduling when the time interval changes [19].

In summary, to avoid the failure of demand defense due to the inability of the current discrete energy storage operation strategy to obtain the state of charge within the section and accurately reflect the energy storage ramping process, as well as the inability of existing continuous-time control strategies to capture dynamic information when the optimization interval length changes, this chapter proposes an energy storage operation strategy based on a continuous-time scale. By deriving interval-normalized Bernstein polynomials to construct a continuous energy storage operation model, a demand defense strategy for light-storage type industrial parks under special scenarios is established, effectively improving the demand defense capability of industrial parks.

This chapter makes the following contributions:

(1) To obtain the continuous energy state and charge–discharge power within any closed interval, the basic properties of interval-normalized Bernstein polynomials are derived based on traditional Bernstein polynomials.
(2) The energy storage model is constructed as a continuous-time scale operation model based on interval-normalized Bernstein polynomials. When there is a

sudden increase in load or a fault occurs, the operation state of energy storage within any time section is accurately captured.

(3) Under the current two-part tariff, an energy storage operation strategy based on a continuous-time scale is proposed to enhance the demand defense capability of light-storage type industrial parks by establishing a day-ahead operation model and an emergency operation model for industrial parks.

(4) Simulation analysis is conducted on a light-storage type industrial park in Zibo, and the results validate that the strategy proposed in this chapter is significantly better than the traditional discrete operation method for energy storage demand defense.

7.2 Renewable Industrial Park

The continuous optimization method based on Bernstein polynomial approximation has been successfully applied to various fields, including power system flexibility improvement, energy storage configuration, and random scheduling [13]. However, traditional Bernstein polynomials are only applicable to the continuous approximation of functions within the unit closed interval $[t, t+1]$. In special scenarios, such as load surges or reductions in photovoltaic power, the interval length may not satisfy this condition. Therefore, the traditional Bernstein polynomial approximation is challenging to apply in the demand management of optical storage parks under these special scenarios.

To address this issue, this chapter proposes an interval-normalized Bernstein polynomial, which extends the application of Bernstein polynomials to continuous optimization scheduling over any closed interval. The definition and related properties of interval-normalized Bernstein polynomials are provided below.

7.2.1 *Definition of Interval Normalized Bernstein Polynomial*

If the function $f_t(x)$ is continuous on $x \in [(t-1)\Delta t, t\Delta t]$, $\forall \Delta t$ let $a = (t-1)\Delta t$, $b = t\Delta t$, then the interval can be converted to a normalized interval by the following equation:

$$x = a(1-r) + br, r \in [0, 1] \tag{7.1}$$

where $f_t(x)$ is continuous on $x \in [a, b]$. The normalized Bernstein polynomial in the interval $x \in [a, b]$ is defined as follows:

$$B_n^x = \sum_{i=0}^{n} f_t\left(a + \frac{i}{n}(b-a)\right) C_n^i \left(\frac{x-a}{b-a}\right)^i \left(\frac{b-x}{b-a}\right)^{n-i} \tag{7.2}$$

where B_n^x denotes the interval-normalized Bernstein polynomial of the continuous function $f_t(x)$, and n is the order of the polynomial. Taking the polynomial function $f_t(a + \frac{i}{n}(b - a))$ as the control point, the continuous approximation curve of the interval normalized Bernstein polynomial on $x \in [a, b]$ can be obtained.

Bernstein polynomial B_n^x based on interval normalization has the property of approximating continuous function $f_t(x)$ on any closed interval. When the order of the interval normalized Bernstein polynomial increases infinitely, the polynomial B_n^x converges uniformly to the continuous function $f_t(x)$ [11].

7.2.2 Derivative and Integral Properties

$$\frac{dB_n^x}{dx} = \sum_{i=0}^{n-1} C_{n-1}^i \left(\frac{x-a}{b-a}\right)\left(_i\frac{b-x}{b-a}\right)^{n-1-i} \frac{n}{b-a} \cdot \left(f_t(a + \frac{i+1}{n}(b-a) - f_t a + \frac{i}{n}(b-a))\right) \tag{7.3}$$

Equation (7.3) demonstrates that, upon derivation, the order of the interval-normalized Bernstein polynomial is reduced from nnn to $n - 1$.

$$\int_a^b B_n^x \mathrm{d}x = \int_a^b \sum_{i=0}^{n} f_t\left(a + \frac{i}{n}(b-a)\right) C_n^i \left(\frac{x-a}{b-a}\right)^i \cdot \left(\frac{b-x}{b-a}\right)^{n-i} \mathrm{d}x = \frac{(b-a)\sum_{i=0}^{n} f_t\left(a + \frac{i}{n}(b-a)\right)}{n+1} \tag{7.4}$$

Equation (7.4) shows that the integral value of the interval normalized Bernstein polynomial depends on the polynomial function $f_t\left(a + \frac{i}{n}(b-a)\right)$ and the order n.

7.2.3 Constraint Reduction Properties

According to the definition of the interval-normalized Bernstein polynomial, the polynomial can be expressed as the product of the following row and column vectors:

$$B_n^x = \sum_{i=0}^{n} f_t\left(a + \frac{i}{n}(b-a)\right) C_n^i \left(\frac{x-a}{b-a}\right)\left(_i\frac{b-x}{b-a}\right)^{n-i}$$

$$
= \left[f_t(a), f_t\left(a + \frac{(b-a)}{n}\right), \ldots, f_t(b) \right] \cdot \begin{bmatrix} C_n^0 (x-a)^0 (b-x)^{n-0} \\ C_n^1 (x-a)^1 (b-x)^{n-1} \\ \cdots \\ C_n^n (x-a)^n (b-x)^{n-n} \end{bmatrix} \left(\frac{1}{b-a} \right)^n \tag{7.5}
$$

From the above equation, it is evident that during the approximation process of interval-normalized Bernstein polynomials, the same column vector can be eliminated from the constraints [14]. Consequently, the equality and inequality constraints can be simplified to involve only polynomial functions.

7.3 Interval Normalized Bernstein Polynomial

7.3.1 Energy Storage Operation with Continuous Time Scale

Energy storage power management during charging and discharging is crucial for achieving peak-valley arbitrage, maximizing local photovoltaic consumption, and enhancing economic benefits in optical storage parks [15]. On a continuous time scale, the kth workshop's energy storage within the park must adhere to the following constraints during the charging and discharging processes:

$$
\begin{cases} -P_{\max}^{\mathrm{ES}} \le P_{t,k}^{\mathrm{ES}}(x) \le P_{\max}^{\mathrm{ES}} \\ \frac{\mathrm{d}S_{t,k}^{\mathrm{soe}}(x)}{\mathrm{d}x} = P_{t,k}^{\mathrm{ES}}(x) \\ S_{\min}^{\mathrm{soc}} \le S_{t,k}^{\mathrm{soc}}(x) = \frac{S_{t,k}^{\mathrm{soe}}(x)}{E_k} \le S_{\max}^{\mathrm{soc}} \\ \int_1^T \left(\int_{(t-1)\Delta t}^{t\Delta t} P_{t,k}^{\mathrm{ES}}(x)\mathrm{d}x \right) \mathrm{d}t = 0 \end{cases} \tag{7.6}
$$

where $x \in [(t-1)\Delta t, t\Delta t]$, $P_{t,k}^{\mathrm{ES}}$ represents the charging and discharging power of the kth workshop energy storage at time t, and $P_{\max}^{\mathrm{ES}}$ is the maximum charging and discharging power of the energy storage. $S_{t,k}^{\mathrm{soe}}$ representing the energy state of the kth workshop energy storage at time t. $S_{t,k}^{\mathrm{soc}}$ denotes the state of charge of the kth workshop energy storage, E_k is the energy storage capacity of the kth workshop, and $S_{\max}^{\mathrm{soc}}$ and $S_{\min}^{\mathrm{soc}}$ are the maximum and minimum SOC values, respectively.

7.3.2 Continuous Time Operation

Based on the properties of interval normalized Bernstein polynomials, the continuous time scale operation model for energy storage can be expressed using Eqs. (7.7) through (7.10). Specifically, by applying Eq. (7.5), the continuous charging and discharging constraints as well as the state of charge constraints for energy storage are equivalent to Eqs. (7.7) and (7.9), respectively. Similarly, Eq. (7.8) represents the energy state constraint of energy storage as derived from Eq. (7.3). Finally, Eq. (7.10) represents the power constraint over one cycle of energy storage, derived from Eq. (7.4).

$$-P_{\max}^{\mathrm{ES}} \le P_{t,k,i}^{\mathrm{ES}} \le P_{\max}^{\mathrm{ES}} \tag{7.7}$$

$$S_{t,k,i}^{\mathrm{soe}} - S_{t,k,i-1}^{\mathrm{soe}} = P_{t,k,i-1}^{\mathrm{ES}} \tag{7.8}$$

$$S_{\min}^{\mathrm{soc}} \le S_{t,k,i}^{\mathrm{soc}} = S_{t,k,i}^{\mathrm{soe}}/E^{k} \le S_{\max}^{\mathrm{soc}} \tag{7.9}$$

$$\Delta t \sum_{t=1}^{T} \sum_{i=0}^{n} P_{t,k,i}^{\mathrm{ES}}/(n+1) = 0 \tag{7.10}$$

where $P_{t,k,i}^{\mathrm{ES}}$ denotes the i-order charging and discharging power of the kth workshop energy storage in t period; $S_{t,k,i}^{\mathrm{soe}}$, $S_{t,k,i}^{\mathrm{soc}}$ are the i-order energy state and state of charge of the kth workshop energy storage in the t-time period, respectively, and Δt represents the time interval of optimal operation.

7.4 Continuous Time Scale Optimization

7.4.1 Day-Ahead Operation Based on Continuous Time Scale

Leveraging the interval normalized Bernstein polynomial, this chapter formulates a day-ahead optimal operation model for the park under a two-part electricity pricing scheme. The primary objective is to minimize the park's electricity cost. The optimization criteria encompass several factors: electricity cost, energy storage operation and maintenance expenses, income from photovoltaic grid integration, and demand-related electricity costs:

$$\min \left[\sum_{t=1}^{T} \sum_{k=1}^{K} \Delta t \sum_{i=0}^{n} \frac{C_t^G P_{t,k,i}^G}{n+1} + C^{\mathrm{ES}} \left| \sum_{i=0}^{n} \frac{P_{t,k,i}^{\mathrm{ES}}}{n+1} \right| - C^{\mathrm{on}} \sum_{i=0}^{n} \frac{P_{t,k,i}^{on}}{n+1} \right] + \frac{C^{\mathrm{DC}} P_0^{G,\max}}{30} \tag{7.11}$$

where the first term represents the electricity tariff paid by the optical storage park to the power grid. Here, K denotes the number of workshops in the park, C_t^G represents the time-of-use electricity price from the power grid, and $P_{t,k,i}^G$ denotes the ith order power purchased by the kth workshop from the grid during time period t. The second term accounts for the operation and maintenance costs of energy storage, with C^{ES} being the operation and maintenance cost coefficient, and $P^{\mathrm{ES}}_{t,k,i}$ representing the ith order charging and discharging power of the kth workshop's energy storage system during period t. The third term reflects the income from photovoltaic grid integration, where C^{on} is the photovoltaic grid price and $P^{\mathrm{on}}_{t,k,i}$ denotes the ith order photovoltaic grid power of the kth workshop during time period t. The fourth term pertains to the daily average demand electricity price, where C^{DC} is the unit power demand price, and $P_0^{\mathrm{G,max}}$ represents the maximum demand defense value for the day-ahead operation.

Utilizing the integral properties of the interval normalized Bernstein polynomial, the maximum demand over the entire operation period of the park is described by Eq. (7.12). Here, the demand power in period t is characterized by the average power drawn from the grid during that period.

$$P_0^{\mathrm{G,max}} = \max\left(\sum_{k=1}^{K}\sum_{i=0}^{n} P^{\mathrm{G}}_{t,k,i}/(n+1)\right) \tag{7.12}$$

During the operation of the park's electrical energy system, power interaction constraints must be met between individual workshops and between the workshops and the power grid. Utilizing the constraint reduction property of the interval normalized Bernstein polynomial, these power interaction constraints can be formulated as follows. Equations (7.13) and (7.14) represent the power interaction constraints between each workshop and the power grid. Equation (7.15) describes the power interaction constraints between different workshops. These formulations ensure that all power interaction requirements are satisfied throughout the operation of the park.

$$0 \le \sum_{k=1}^{K} P^{G}_{t,k,i} \le P^{G}_{\max} \tag{7.13}$$

$$P^{\mathrm{on}}_{t,k,i} \ge 0 \tag{7.14}$$

$$P^{\mathrm{in}}_{t,k,i} = \eta^2 P^{\mathrm{out}}_{t,k,i},\ P^{\mathrm{in}}_{t,k,i} \ge 0,\ P^{\mathrm{out}}_{t,k,i} \ge 0 \tag{7.15}$$

During the operation of the optical storage park, the power balance constraints for each workshop are specified as follows:

$$P^{\mathrm{ES}}_{t,k,i} - P^{\mathrm{in}}_{t,k,i} + P^{\mathrm{on}}_{t,k,i} + P^{\mathrm{out}}_{t,k,i} - \eta P^{G}_{t,k,i} - P^{V}_{t,k,i} + P^{d}_{t,k,i} = 0 \tag{7.16}$$

To ensure the continuous operation of the optical storage park, the n-order power of the $t-1$ period must be equal to the 0-order power of the t period.

$$P^{\mathrm{ES}}_{t-1,k,n} = P^{\mathrm{ES}}_{t,k,0},\ P^{G}_{t-1,k,n} = P^{G}_{t,k,0},\ P^{\mathrm{on}}_{t-1,k,n} = P^{\mathrm{on}}_{t,k,0}$$

$$P^{\text{in}}_{t-1,k,n} = P^{\text{in}}_{t,k,0}, P^{\text{out}}_{t-1,k,n} = P^{\text{out}}_{t,k,0} \tag{7.17}$$

Under normal operating conditions, the demand defense model of the optical storage park is defined by Eqs. (7.7) through (7.17). Solving this mathematical model enhances the park's demand defense capabilities while ensuring its economic operation.

7.4.2 Emergency Operation Model Based on Continuous Time Scale

Under special scenarios (heavy rain, hail, or sudden start of high-power equipment, etc.), the sudden reduction or load surge of photovoltaic power generation can easily lead to the failure of demand defense in the park, and it is necessary to make emergency decisions in time. On the basis of introducing the normal operation of the park, this section gives the emergency operation strategy of the park under special scenarios.

(1) Normal operation model based on continuous time scale

In Fig. 7.1, x_b denotes the moment a special event occurs, leading to a modification in the time intervals. Specifically, the day is divided into 97 time periods, each with a 15-min interval ($\Delta t = 15$ min), rather than the usual 96 periods. The interval $[1, t_c]$ represents the time period prior to the special event, while $[t_h, 97]$ indicates the periods following the event. Each interval's length is denoted by $L_t = [\Delta t, \Delta t, \ldots, x_b - x_1, x_2 - x_b, \ldots, \Delta t]$.

Before initiating emergency operations, it is essential to calculate the electricity cost of the park for the periods preceding t_c, as outlined in Eqs. (7.18)–(7.21):

$$C^n = \sum_{t=1}^{t_c-1}\sum_{k=1}^{K} \Delta t \left[\sum_{i=0}^{n} \frac{C^G_t P^G_{t,k,i}}{n+1} + C^{\text{ES}} \left| \sum_{i=0}^{n} \frac{P^{\text{ES}}_{t,k,i}}{n+1} \right| - C^{\text{on}} \sum_{i=0}^{n} \frac{P^{\text{on}}_{t,k,i}}{n+1} \right]$$

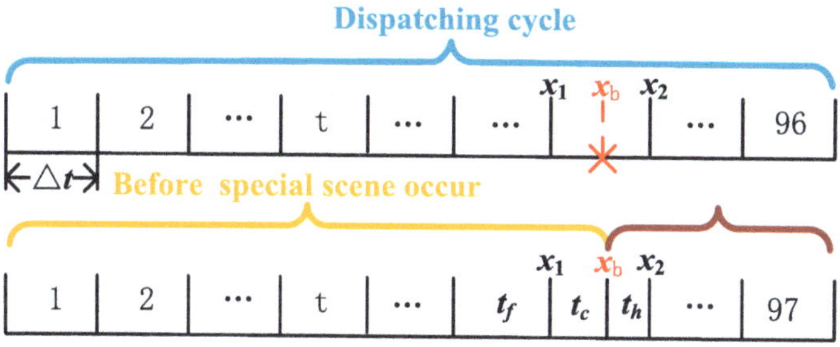

Fig. 7.1 Schematic diagram of the occurrence time of extreme scenario

$$+\sum_{k=1}^{K}\left[C_{t_c}^{G}S_k^{G}+C^{\text{on}}S_k^{\text{on}}+C^{\text{ES}}|S_k^{\text{ES}}|\right] \tag{7.18}$$

where C^n denotes the electricity cost of the park before the occurrence of special scenarios; S_k^G, S_k^{on}, and S_k^{ES} are the kth workshop's purchasing electricity from the grid, photovoltaic power, and energy storage charge and discharge before the t_c period, respectively. These variables are determined using by Eqs. (7.19)–(7.21).

$$S_k^G=\int_{x_1}^{x_b}\sum_{i=0}^{n}P_{t_c,k,i}^{G}C_n^i\left(\frac{x-x_1}{b-a}\right)\left(_i\frac{x_2-x}{b-a}\right)^{n-i}\mathrm{d}x \tag{7.19}$$

$$S_k^{\text{on}}=\int_{x_1}^{x_b}\sum_{i=0}^{n}P_{t_c,k,i}^{\text{on}}C_n^i\left(\frac{x-x_1}{b-a}\right)^i\left(\frac{x_2-x}{b-a}\right)^{n-i}\mathrm{d}x \tag{7.20}$$

$$S_k^{\text{ES}}=\int_{x_1}^{x_b}\sum_{i=0}^{n}P_{t_c,k,i}^{\text{ES}}C_n^i\left(\frac{x-x_1}{b-a}\right)^i\left(\frac{x_2-x}{b-a}\right)^{n-i}\mathrm{d}x \tag{7.21}$$

(2) Emergency operation model based on continuous time

The emergency operation is designed to address the demand defense challenges arising from a sudden drop in photovoltaic power generation or a surge in load, by utilizing energy storage in the park. The objective function encompasses several cost components: the electricity cost, the operation and maintenance cost of energy storage, the income from photovoltaic grid connection, the demand electricity cost, and the electricity cost incurred by the park before the occurrence of the special event.

$$\min\sum_{t=t_h}^{T+1}\sum_{k=1}^{K}L_t\left[\sum_{i=0}^{n}\frac{C_t^G P_{t,k,i}^G}{n+1}+C^{\text{ES}}\left|\sum_{i=0}^{n}\frac{P_{t,k,i}^{\text{ES}}}{n+1}\right|+C^{\text{on}}\sum_{i=0}^{n}\frac{P_{t,k,i}^{\text{on}}}{n+1}\right]+\frac{C^{\text{DC}}P_0^{G,\max}}{30}+C^n \tag{7.22}$$

$$\begin{cases}P^{G,\max}\geq\max\left(\sum\limits_{k=1}^{K}\sum\limits_{i=0}^{n}P_{t,k,i}^{G}/(n+1)\right),t\in[t_h+1,97]\\ P^{G,\max}\geq\left(\sum\limits_{k=1}^{K}\sum\limits_{i=0}^{n}L_{t_h}P_{t_h,k,i}^{G}/(n+1)+\sum\limits_{k=1}^{K}S_k^{G}\right)/\Delta t\\ P^{G,\max}\geq P_0^{G,\max}\end{cases} \tag{7.23}$$

where $P^{G,\max}$ denotes the maximum demand for the entire operation cycle of the park. According to Eq. (7.23), the maximum demand during special scenarios must meet or exceed the demand defense value established under normal operation conditions.

In the event of a sharp decline in photovoltaic output or a significant increase in load, the power balance constraints for each workshop are as follows:

$$P_{t,k,i}^{\mathrm{ES}} + P_{t,k,i}^{\mathrm{on}} + P_{t,k,i}^{\mathrm{out}} + \beta P_{t,k,i}^{d} = \eta P_{t,k,i}^{G} + \alpha P_{t,k,i}^{V} + P_{t,k,i}^{\mathrm{in}} \tag{7.24}$$

where α and β represent the coefficients for photovoltaic output and load demand, respectively, in special scenarios.

Based on the definition of the interval normalized Bernstein polynomial, the initial values for the emergency operation can be computed using Eqs. (7.25) through (7.29).

$$P_{t_h,k,0}^{\mathrm{ES}} = \sum_{i=0}^{n} P_{t_c,k,i}^{\mathrm{ES}} C_n^i \left(\frac{x_b - x_1}{b-a}\right)^i \left(\frac{x_2 - x_b}{b-a}\right)^{n-i} \tag{7.25}$$

$$P_{t_h,k,0}^{\mathrm{on}} = \sum_{i=0}^{n} P_{t_c,k,i}^{\mathrm{on}} C_n^i \left(\frac{x_b - x_1}{b-a}\right)^i \left(\frac{x_2 - x_b}{b-a}\right)^{n-i} \tag{7.26}$$

$$P_{t_h,k,0}^{\mathrm{in}} = \sum_{i=0}^{n} P_{t_c,k,i}^{\mathrm{in}} C_n^i \left(\frac{x_b - x_1}{b-a}\right)^i \left(\frac{x_2 - x_b}{b-a}\right)^{n-i} \tag{7.27}$$

$$P_{t_h,k,0}^{\mathrm{out}} = \sum_{i=0}^{n} P_{t_c,k,i}^{\mathrm{out}} C_n^i \left(\frac{x_b - x_1}{b-a}\right)^i \left(\frac{x_2 - x_b}{b-a}\right)^{n-i} \tag{7.28}$$

$$S_{t_h,k,0}^{\mathrm{soe}} = \sum_{i=0}^{n} S_{t_c,k,i}^{\mathrm{soe}} C_n^i \left(\frac{x_b - x_1}{b-a}\right)^i \left(\frac{x_2 - x_b}{b-a}\right)^{n-i} \tag{7.29}$$

In summary, the demand defense model for continuous energy storage operation in the park, as proposed in this chapter, is composed of Eqs. (7.7)–(7.10), (7.13)–(7.15), and (7.17)–(7.29). Solving this mathematical model ensures both the economic operation of the park and enhanced demand defense capability during special scenarios.

7.5 Case Study

This chapter presents a simulation and analysis of a photovoltaic energy storage park located in Zibo. The park's structure is depicted in Fig. 7.2. It comprises two workshops, each equipped with a photovoltaic energy storage power station. The transformer loss is set at 0.02, and the energy storage charge and discharge efficiency is 90%.

To address practical engineering needs and facilitate comparison with traditional discrete models, the day is divided into 96 periods, each lasting 15 min. The 5th-order interval normalized Bernstein polynomial is employed for continuous approximation of the park, and the Gurobi solver is utilized to solve the problem with an error margin of 0.1%.

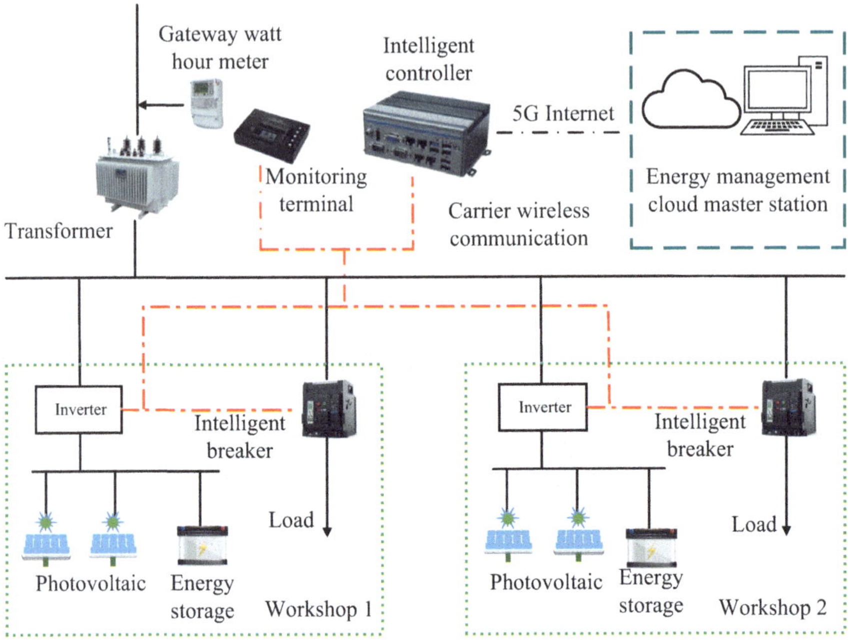

Fig. 7.2 System structure of a photovoltaic and energy storage integrated park in Zibo city

Prior to the occurrence of any special scenarios, each workshop implements a day-ahead operational strategy. This chapter compares the traditional discrete optimization approach with a continuous optimization strategy based on interval normalized Bernstein polynomials for optimizing the day-ahead operation model of the photovoltaic energy storage park. Figure 7.3 illustrates the day-ahead optimal operation curves for both Workshop 1 and Workshop 2 under these different operational strategies.

In the analysis, DT refers to the traditional discrete optimization strategy utilizing 96 operating points, whereas CT denotes the continuous time scale energy storage operation strategy proposed in this chapter to enhance the demand defense capability of the park. Notably, the load and interactive power under the CT operation strategy, as depicted in Fig. 7.3, are derived from the continuous time model outlined in Eqs. (7.25)–(7.29), with the total actual load equating to the mean value.

Figure 7.3 demonstrates that, under both CT and DT strategies, energy storage is optimized by charging during periods of low grid prices or high photovoltaic output and discharging when grid prices are high or photovoltaic output is low, thus facilitating peak-valley arbitrage. Additionally, due to its relatively lower load, Workshop 2 transmits excess electric energy to Workshop 1 during periods of abundant energy to maximize the local utilization of photovoltaic resources. Consequently, Workshop

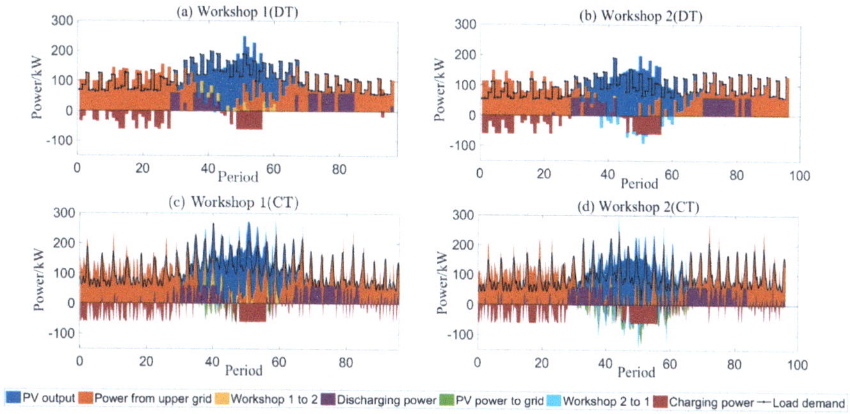

Fig. 7.3 Day-ahead operation curves of photovoltaic and energy storage integrated park under CT and DT strategies

2 exhibits higher online and interactive power compared to Workshop 1. Under the CT model, despite a significant power spike in Workshop 2 between periods 40 and 50, it remains within the maximum demand value for 15-min intervals, thereby avoiding additional demand costs.

To assess the approximation accuracy of the Bernstein polynomial-based CT operation strategy, Table 7.1 presents a comparison of electricity costs in the optical storage park under both DT and CT operation strategies. As shown in Table 7.1, the day-ahead electricity costs for the park under the CT operation strategy closely align with those under the DT operation strategy. This indicates that the interval normalized Bernstein polynomials offer a reliable approximation accuracy.

Compared to DT operation strategy, the CT operation strategy more effectively captures the ongoing state of energy storage at any given moment, providing a precise representation of the energy storage's operational dynamics. The DT strategy, by contrast, only assesses the energy storage operation at specific time points. As a result, sudden changes in the park's operational state within a discrete time period

Table 7.1 Operating costs of photovoltaic storage park under CT and DT strategies

Operation strategy	Electricity tariff/ CNY	Internet revenue/ CNY	Energy storage operation and maintenance/ CNY	Electricity demand tariff/ CNY
DT	2057.7	0.0	68.2	279.3
CT	2057.4	0.0	68.3	279.3
Error	0.015%	0	0.15%	0

may not be accurately reflected by the DT strategy, potentially impairing its ability to respond to emergencies. This limitation can compromise the park's demand defense capability and affect the overall profitability of the optical storage power station.

The special scenarios for the park include abrupt reductions in photovoltaic power generation or sudden surges in load due to high-power equipment (especially experimental machinery) during extreme weather conditions. The timing and magnitude of such shortages cannot be accurately predicted in advance. In these scenarios, the interaction power between the park and the power grid can increase significantly, potentially leading to a failure in meeting the monthly demand defense and a rise in monthly electricity costs. Assuming the special scenario lasts for 15 min, this chapter examines a 10% reduction in photovoltaic power generation as the case study.

Figure 7.4 illustrates the park's maximum demand based on both DT and CT operation strategies when the special scenario occurs at different times. The figure reveals that if the special scenario occurs during the period from 12:30 to 12:45 (the 51st period of the day-ahead operation), both DT and CT strategies can effectively manage the surge in demand by adjusting the energy storage charging and discharging power. However, if the special scenario occurs between 12:33 and 12:48 or between 12:36 and 12:51, overlapping with the transition between the 51st and 52nd periods, the DT strategy fails to capture the energy storage state during these periods. Consequently, power shortages are compensated by the grid, leading to an increase in the park's maximum demand. Conversely, the CT strategy, by leveraging continuous time modeling, can formulate an appropriate emergency response based on the energy storage state at the time of the special scenario, thereby effectively managing the demand surge.

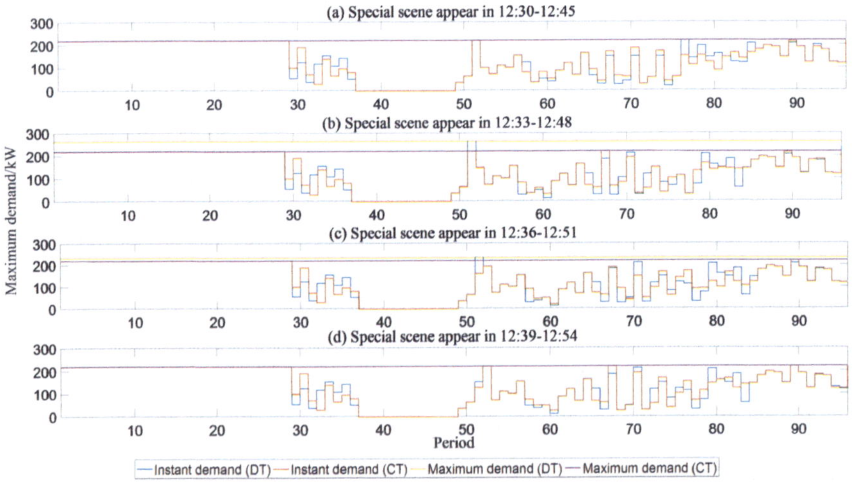

Fig. 7.4 DT and CT operation strategies when the special scenario occurs at different times

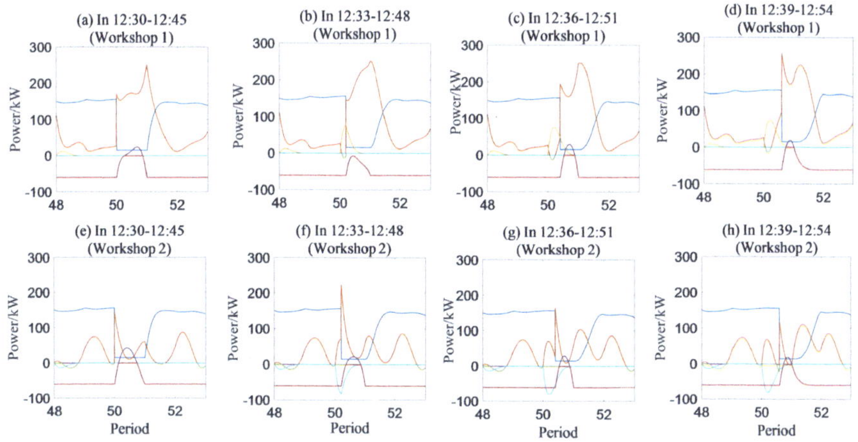

Fig. 7.5 Emergency response of power park based on CT operation strategy when extreme scenario occurs

When the special scenario occurs between 12:39 and 12:54, which overlaps the transition between periods 51 (12:30–12:45) and 52 (12:45–13:00), the DT operation strategy relies on power from the higher grid to address the park's power shortage. Notably, due to the brief duration of the special scenario within the 51st period, the state of charge of the energy storage has minimal impact on the DT strategy. As a result, the grid-supplied power during this scenario does not exceed the maximum demand defense threshold. This indicates that when a special scenario arises during the latter part of a given day, both DT and CT operation strategies can effectively manage the park's demand without surpassing the maximum demand defense value.

To further evaluate the impact of the CT operation strategy on the park's emergency response, Fig. 7.5 presents a comparison of the emergency responses under the CT operation strategy at various times.

From Fig. 7.5, it is evident that when a special scenario occurs with a sharp decline in photovoltaic output, both Workshop 1 and Workshop 2 implement effective emergency measures to prevent a significant increase in maximum demand. Specifically, Workshop 1 promptly shifts the energy storage from charging to discharging, mitigating the power shortage and preventing a rise in maximum demand. Similarly, Workshop 2 adjusts its strategy by not only transitioning its energy storage from charging to discharging but also by reducing the power exchange with Workshop 1 (transferring power to Workshop 1). This approach helps in avoiding an increase in the maximum demand.

Table 7.2 compares the electricity costs and demand charges in the power park using CT and DT operation strategies at different times during special scenarios. As illustrated in Table 7.2, when the special scenario occurs between 12:30 and 12:45, both DT and CT operation strategies result in similar total electricity costs and demand charges. However, at other times, the CT operation strategy demonstrates superior demand defense capabilities compared to the DT strategy, reducing

Table 7.2 Cost comparison of park under CT and DT strategies

Operation strategy	Special scene moments	Total cost/CNY	Electricity demand tariff/CNY
DT	12:30–12:45	2448.1	279.3
	12:33–12:48	2495.6	338.1
	12:36–12:51	2456.6	299.4
	12:39–12:54	2435.3	279.3
CT	12:30–12:45	2447.5	279.3
	12:33–12:48	2434.9	279.3
	12:36–12:51	2431.5	279.3
	12:39–12:54	2429.2	279.3

electricity demand by approximately 0–20%. Specifically, the closer the occurrence time of the special scenario is to the start of the period, the greater the increase in maximum demand observed with the DT strategy. Conversely, when the special scenario occurs later in the period and the DT strategy does not exceed the demand defense threshold, the demand charges remain unchanged.

In summary, this chapter addresses the issue of sudden drops in photovoltaic power generation during special scenarios by proposing a continuous operation strategy based on interval normalized Bernstein polynomials. This strategy effectively manages the charging and discharging of energy storage in real-time, mitigating increases in the park's maximum demand and enhancing its demand defense capabilities. It offers a practical solution for maintaining the economic operation of the photovoltaic energy storage park under such challenging conditions.

7.6 Conclusion

(1) This chapter proposes a continuous time-scale energy storage operation strategy to enhance the demand defense capabilities of photovoltaic energy storage parks. This strategy enables real-time assessment of the energy storage state and accurately captures the energy storage dynamics throughout any given time interval.

(2) In comparison with the traditional DT operation strategy, the CT strategy based on interval normalized Bernstein polynomials introduced in this chapter significantly improves the demand defense capability of the photovoltaic energy storage park while maintaining high approximation accuracy. This CT approach achieves a 100% success rate in demand defense and can reduce the maximum demand electricity cost by up to 20%. It offers an innovative solution for managing demand defense during special scenarios.

(3) The proposed method addresses the limitation of traditional Bernstein polynomials, which are only applicable for continuous optimization within a unit interval. By extending Bernstein polynomials to any closed interval, this method enhances their applicability to continuous optimization scheduling. It holds potential for improving power system resilience and flexibility, optimizing energy storage configurations, and addressing random scheduling challenges in various fields.

References

1. Xie X, Ma Ningjia, Liu W et al (2022) Functions of energy storage in renewable energy dominated power systems: review and prospect. Proc CSEE 1–12
2. Huang R, Zhang S, Wang P (2022) Key areas and pathways for carbon emissions reduction in Beijing for the "Dual Carbon" targets. Energy Policy 164:112873
3. National Development and Reform Commission, National Energy Administration. Notice on developing new types of energy storage [EB/OL] (2021)
4. Wang Y, Liu Z, Cai C et al (2022) Research on the optimization method of integrated energy system operation with multi-subject game. Energy 245:123305
5. Wang YX, Chen JJ, Zhao YL et al (2024) Incorporate robust optimization and demand defense for optimal planning of shared rental energy storage in multi-user industrial park. Energy 301:131721
6. Cucchiella F, D'adamo I, Gastaldi M (2016) Photovoltaic energy systems with battery storage for residential areas: an economic analysis. J Clean Prod 131:460–474
7. Jing Y, Zhu L, Yin B et al (2023) Evaluating the PV system expansion potential of existing integrated energy parks: a case study in North China. Appl Energy 330:120310
8. Cao D, Zhao J, Hu W et al (2021) Robust deep Gaussian process-based probabilistic electrical load forecasting against anomalous events. IEEE Trans Ind Inf 18(2):1142–1153
9. Yan GJ, Chen JJ, Liu JY et al (2023) Random clustering and dynamic recognition-based operation strategy for energy storage system in industrial park. J Energy Storage 73:109192
10. Khatami R, Parvania M (2019) Stochastic multi-fidelity scheduling of flexibility reserve for energy storage. IEEE Trans Sustain Energy 11(3):1438–1450
11. Zheng C, Fang J, Wang S et al (2020) Energy flow optimization of integrated gas and power systems in continuous time and space. IEEE Trans Smart Grid 12(3):2611–2624
12. Nikoobakht A, Aghaei J, Shafie-Khah M et al (2020) Continuous-time co-operation of integrated electricity and natural gas systems with responsive demands under wind power generation uncertainty. IEEE Trans Smart Grid 11(4):3156–3170
13. Wu D, Han S, Wang L et al (2023) Multi-parameter optimization design method for energy system in low-carbon park with integrated hybrid energy storage. Energy Convers Manage 291:117265
14. Khatami R, Parvania M, Khargonekar PP (2017) Scheduling and pricing of energy generation and storage in power systems. IEEE Trans Power Syst 33(4):4308–4322
15. Kim YS, Park GH, Kim SW et al (2024) Incentive design for hybrid energy storage system investment to PV owners considering value of grid services. Appl Energy 373:123772
16. Ahmad F, Ashraf I, Iqbal A et al (2024) Energy management strategies and cost benefits analysis at electric vehicle parking lots incorporating photovoltaic energy generation and energy storage system. J Energy Storage 92:112105
17. Khatami R, Parvania M, Narayan A (2019) Flexibility reserve in power systems: definition and stochastic multi-fidelity optimization. IEEE Trans Smart Grid 11(1):644–654

18. Zhou B, Fang J, Ai X et al (2021) Flexibility-enhanced continuous-time scheduling of power system under wind uncertainties. IEEE Trans Sustain Energy 12(4):2306–2320
19. Yang W, Huang Y, Ye T et al (2024) Hybrid continuous-discrete time control strategy to optimize thermal network dynamic storage and heat-electricity integrated energy system dynamic operation. Energy Convers Manage 301:118035

Part III
Energy Storage System Optimization

Chapter 8
Random Clustering and Dynamic Recognition Strategy for Energy Storage System Optimization

8.1 Introduction

With the continual widening of the peak-valley price differential and the rapid advancements in storage technology, Energy Storage Systems (ESS) have emerged as pivotal elements in enhancing the economic viability of industrial parks [1–3]. On one hand, it aids in narrowing the gap between peak and off-peak load power, thereby reducing the costs associated with demand tariffs linked to transformer capacity under the two-part tariff structure [4]. This capacity to level out load profiles is particularly beneficial for industrial parks, optimizing their energy consumption patterns.

Furthermore, ESS enables local integration and utilization of Photovoltaic (PV) power generation within industrial park settings [5]. This localized energy storage capacity not only enhances the reliability of renewable energy integration but also contributes to the overall efficiency and sustainability of the park's energy ecosystem. Moreover, the deployment of ESS within industrial parks offers additional advantages. It can delay the need for costly upgrades to power infrastructure and alleviate potential grid congestion, thus ensuring smooth and efficient operation of the utility network [6]. As highlighted in [7], the declining prices of ESS technologies underscore their potential as a future-oriented development trend for industrial parks. This trend underscores the increasing importance of devising optimal operation strategies for ESS within industrial park settings.

In recent years, the application of ESS in industrial parks has raised substantial attention from researchers and practitioners alike [9, 10]. The focus lies on developing strategies that maximize the economic benefits of ESS deployment while ensuring reliable and sustainable energy supply for industrial operations. This surge of interest underscores the pivotal role that ESS plays in the evolving landscape of industrial park energy management and its potential to significantly impact operational efficiency and cost savings.

J. Chen and Y. Li, *Renewable Power System Optimization*, Smart Energy Systems,
https://doi.org/10.1007/978-981-97-8132-4_8

The researches on ESS operation within industrial parks are generally classified into two main categories. The first category concentrates on the deterministic optimization. For instance, in [10], researchers explored the ESS optimal operation within a PV-integrated industrial park, aiming to minimize operational costs. Similarly, [11] delved into the optimal operation of ESS, with objectives centered on minimizing carbon emissions and operating costs. However, these studies often overlook the random fluctuations, leading to suboptimal operation strategies [12].

In light of these limitations, the second category focusing on uncertain optimization has garnered significant attention in the past few years. Methods derived from this line of research primarily fall into two categories: stochastic and robust optimizations [13, 14]. These approaches are designed to address the inherent uncertainties in industrial park operations, offering more robust and adaptable strategies for ESS utilization.

Stochastic optimization methods take into account the probabilistic nature of uncertainties, allowing for the incorporation of random variables such as fluctuating power demand or renewable energy generation. These methods aim to optimize ESS operation strategies by considering various scenarios and their associated probabilities [15]. On the other hand, robust optimization methods focus on developing strategies that are resilient to a wide range of uncertainties. These approaches aim to find solutions that perform well across different scenarios, even under the most adverse conditions [16].

Robust optimization stands out as a prominent method for handling uncertain parameters, where the range of fluctuations is expressed by an uncertain set, without the need to specify distributions for uncertain parameters [17]. This optimization often leans toward worst-case scenarios, yielding overly conservative that may result in diminished economic efficiency and higher operating costs. In contrast, stochastic optimization emerges as a powerful manner for analyzing and modeling intricate decision problems within power systems, especially those entailing uncertain parameters [18, 19]. By allowing the incorporation of probability distributions to depict uncertainty, stochastic optimization offers more adaptable solutions [20].

For example, in [21], a chance-constrained optimization was developed to address uncertainties related to renewable energy and load demand. While stochastic optimization yields cost-effective results, accurately describing the probability distributions of uncertainty necessitates a substantial number of repeated samples. This inevitably escalates computational complexity, rendering it less suitable for real-world applications within industrial parks.

From a risk aversion perspective, in [22], the metrics of Expected Power Not Served (EPNS) and Expected Power Curtailment (EPC) were introduced to quantify the uncertain risk associated with wind power within stochastic agricultural microgrid scheduling. Similarly, in [23], EPNS and EPC indicators were derived from a risk aversion standpoint for uncertain renewable energy. However, it is noteworthy that the nonlinear expressions of EPNS and EPC derived in [22] and [23] suffer from the curse of dimensionality that often affects stochastic and dynamic programming approaches.

This limitation necessitates careful consideration and potentially the exploration of alternative methodologies to mitigate computational complexity while ensuring accurate representation of uncertainties in ESS optimization within industrial park contexts.

On one hand, load curves within industrial parks can exhibit significant variations, even within the span of a single working day [24]. Developing or adjusting the ESS operation on a daily time for industrial parks may not be cost-effective, primarily due to the associated communication costs. Communication costs remain a principal constraint compared to other factors [25, 26]. Additionally, the communication errors further diminish the necessity of daily updates to the ESS operation [27].

Hence, it becomes imperative to delve into the analysis of electricity consumption behavior and craft a cost-effective ESS operation. Data mining technology has emerged as a potent tool for probing into users' consumption behavior [28]. Notably, clustering, an unsupervised data mining, has been leveraged for this purpose in various studies [29].

By employing clustering methods, the distinctive consumption characteristics are categorized into different clusters. Users within the same cluster tend to exhibit similar consumption patterns [30]. This clustering approach offers a streamlined path to understanding and managing the intricacies of industrial park electricity consumption behavior.

For example, in [31], clustering techniques are utilized to address the inherent uncertainty associated with wind power. Similarly, [32] proposes a multi-stage and multi-objective energy management tailored to navigate the challenges of microgrid scheduling amidst uncertainties.

Furthermore, the uncertainty dilemma outlined in [33] is deftly tackled through the effective application of the k-means data clustering method. This approach not only optimizes the operational efficiency of ESS within industrial parks but also minimizes the need for frequent adjustments, ensuring a cost-effective and robust energy management system.

Within the context outlined above, the clustering technique for analyzing electricity consumption in the context of ESS operation offers a fresh perspective. As the field of clustering, a critical issue arises regarding the determination of the optimal number of clusters. It is crucial to note that in [31–33], the evaluation of the clustering effect primarily revolves around the compactness within the load clusters and the separation between them. However, these evaluations often overlook the balance of economic benefits and uncertain cost inherent in the load clustering.

The determination of the optimal number of clusters holds significant implications for the effectiveness and efficiency of the resulting ESS operation strategy. This decision directly impacts the granularity with which the electricity consumption patterns are categorized, consequently influencing the precision of the ESS operation adjustments. In practical scenarios, the optimal number of clusters cannot always be discerned a priori due to the dynamic and complex nature of industrial park operations. Therefore, methodologies that address this challenge and incorporate economic considerations and load uncertainty into the clustering process are essential for enhancing the efficacy of ESS operation.

From the above literature review, several key observations emerge: (1) The prevailing literature on optimal clustering overlooks considerations of load uncertainty and fails to incorporate the tangible costs inherent to industrial parks. (2) The existing body of literature on uncertain analysis tends to exhibit an overly conservative approach, potentially leading to higher operation costs and diminished economic efficiency. Alternatively, some methods require a substantial number of repeated samplings, thereby decreasing computational efficiency. (3) Despite the maturity of research on multi-feature recognition in various fields, scant attention has been directed towards the exploration of load recognition.

In light of these observations, the main contributions of this chapter can be highlighted as follows:

1. A novel random clustering is proposed to investigate the optimal number of clusters for load curves within industrial parks. This method addresses the limitations of existing clustering approaches by explicitly considering load uncertainty and economic benefits.
2. The linearized EPNS and EPC measures are derived. These indexes serve as essential tools for assessing uncertain risk, achieved through relaxation and piecewise linearization techniques.
3. A load curve dynamic recognition technology is introduced, grounded in the analysis of various characteristics. This innovative approach enables the recognition of the cluster to which an intra-day load curve belongs, subsequently providing the corresponding ESS operation tailored to the specific needs of an industrial park.

By integrating these advancements, this research aims to offer a comprehensive framework for optimizing ESS operation strategies within industrial parks. Through the incorporation of load uncertainty considerations, economic efficiency assessments, and dynamic load curve recognition, this methodology seeks to enhance the operational resilience and cost-effectiveness of industrial park energy management systems.

The rest of the chapter is organized as follows. Section 8.2 presents the random clustering technique. Section 8.3 shows details of the developed dynamic recognition method. Section 8.4 gives the stochastic optimization model with random clustering and dynamic recognition. Finally, numerical case studies are conducted in Sect. 8.5, and conclusions are drawn in Sect. 8.6.

8.2 Random Clustering Technique

Figure 8.1 illustrates the diagram of an industrial park, which includes PV power, load, and ESS. The cost considerations for the industrial park encompass electricity tariffs and demand tariffs within a two-part pricing. The demand tariff is

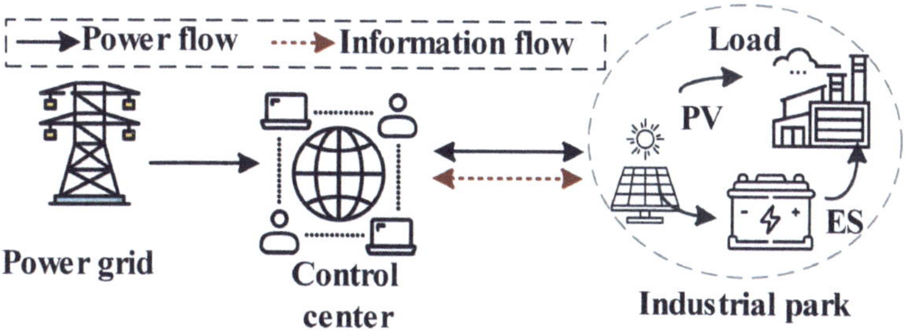

Fig. 8.1 Schematic diagram of industrial park supply system

applied based on the maximum demand recorded each month and multiplied by the corresponding demand price. The electricity tariff is determined by multiplying the real-time electricity consumption by the time-of-use electricity price.

The optimization shown in Eq. (8.1), focuses on the ESS operation for the industrial park. The primary goal, in line with established economic principles [29], is to minimize the overall system operational cost. This cost framework includes power tariffs, demand tariffs, and on-grid tariffs. The optimization is governed by both equality and inequality constraints. Equality constraints address power balance and ESS state constraints. Inequality constraints define the limits on energy storage charge and discharge rates, as well as SOC constraints.

$$\begin{aligned} \text{obj} \quad & F_0 = \min \sum_{g \in G} \sum_{t \in T} \left(P_{t,g}^{\text{buy}}(u)\rho_t - P_{t,g}^{\text{sell}}(u)\rho^{\text{TG}} \right) \\ & \qquad + \rho^{\text{DP}} \max_{t \in T, g \in G} \left\{ P_{t,g}^{\text{buy}}(u) \right\} \\ \text{s.t.} \quad & h(u) = 0 \\ & v(u) \geq 0 \end{aligned} \tag{8.1}$$

where F_0 and u represent the objective function and decision variables, respectively. $G = \{1, 2, \ldots, 30\}$ and $T = \{1, 2, \ldots, 24\}$ are, respectively, considered days and hours; ρ_t, ρ^{TG}, ρ^{DP} indicate time-of-use price, on-grid price, and demand price, respectively; $P_{t,g}^{\text{buy}}$ and $P_{t,g}^{\text{sell}}$ refer to purchasing/selling power from/to upper grid at time t of day g; and $P_{t,g}^{\text{load}}$ illustrates the load power at time t of day g.

Given the certainty of load conditions, solving Eq. (8.1) provides the optimal ESS operation. However, load curves exhibit significant variability across different days. Iteratively adjusting the ESS operation strategy daily introduces economic inefficiencies, primarily due to the associated communication costs [30]. Additionally, frequent adjustments are redundant and may be further complicated by potential communication errors within the industrial park environment [31]. To address these issues, we propose a novel random clustering technique. This approach effectively determines the optimal number of clusters for the industrial park's load curves and identifies the corresponding optimal ESS operation curves. This method addresses the limitations of existing clustering techniques, which often overlook the uncertainties in load variations and the economic benefits within the practical constraints of real-world operations.

8.3 Dynamic Recognition Method

Considering the inherent uncertainties in load, a risk-averse approach is employed to derive the EPNS and EPC measures relevant to uncertain load demand. To facilitate analysis and calculation, these indices are converted into linear expressions through relaxation and piecewise linearization techniques. The optimal number of clusters for the park's load curve can be achieved by using the SCE method, focusing on minimizing expected costs. The optimization model involves the following four steps:

Step 1: Derive the load's probability density function assuming a Gaussian distribution. Calculate the EPNS and EPC measures. These measures are then converted into linear expressions via relaxation and piecewise linearization to simplify analysis.

Step 2: Cluster the load data, setting an objective to minimize the total electricity cost over a one-month period. Optimize the ESS operation for each possible number of clusters.

Step 3: Evaluate the clustering by considering two critical factors: the cost error due to clustering effects and the uncertainty risk associated with operational costs. Determine the optimal number of clusters for the park's load curve using Eq. (8.11) within the SCE framework.

Step 4: Utilize K ESS operation curves corresponding to the optimal cluster count identified in the previous step. These profiles will serve as the ESS operation, finalizing the proposed framework effectively.

8.3.1 *Linearized EPC and EPNS for Uncertain Load*

Assuming that the number of clusters is K, and each cluster's error follows a normal distribution, in the kth cluster, the power load at time t can be represented as follows:

$$P_{t,g}^{\text{load}} \sim N\left(\mu_{t,k}, \sigma_{t,k}^2\right) \tag{8.2}$$

where $g \in G_k \cap G$, G_k indicates the set of load curves belonging to the k th cluster; $\mu_{t,k}$ and $\sigma_{t,k}$ denote the mean and standard deviation of load curves in G_k at time t, respectively.

The probability density function (PDF) $f\left(P_{t,k}^{\text{load}}\right)$ of the load can be expressed as:

$$f\left(P_{t,g}^{\text{load}}\right) = \frac{1}{\sigma_{t,k}\sqrt{2\pi}} e^{-\frac{\left(P_{t,g}^{\text{load}} - \mu_{t,k}\right)^2}{2\left(\sigma_{t,k}\right)^2}} \tag{8.3}$$

If the confidence level for power load is set at 95%, it implies that there is a 5% probability that the load will fall outside the interval, which corresponds to the significance level α. That is, $P_{t,g}^{\text{load}} \in \left[\mu_{t,k} - 1.96\sigma_{t,k}, \mu_{t,k} + 1.96\sigma_{t,k}\right]$. Thus, the minimum and maximum for cluster k can be expressed as $P_{\min^{L,k}} = \mu_{t,k} - 1.96\sigma_{t,k}$, $P_{\max^{L,k}} = \mu_{t,k} - 1.96\sigma_{t,k}$.

The EPC $P_{t,k}^{\text{EPC}}$ and EPNS $P_{t,k}^{\text{EPNS}}$ for the uncertain load $P_{t,k}^{\text{load}}$ at time t for the kth cluster are formulated as Eqs. (8.4) and (8.5), respectively:

$$P_{t,k}^{\text{EPC}} = \int_{P_{t,k}^{\text{load}}}^{P_{\max^{L,k}}} \left(x - P_{t,k}^{\text{load}}\right) f(x) dx \tag{8.4}$$

$$P_{t,k}^{\text{EPNS}} = \int_{P_{\min^{L,k}}}^{P_{t,k}\text{load}} \left(P_{t,k}^{\text{load}} - x\right) f(x) dx \tag{8.5}$$

To enhance solution efficiency, it is crucial to linearize $P_{t,k}^{\text{EPC}}$ and $P_{t,k}^{\text{EPNS}}$, as outlined in Eqs. (8.4) and (8.5), respectively. Prior to this, it is necessary to demonstrate that the terms on the right side of these equations are convex. A sufficient condition for establishing this property is that the second-order derivatives are non-negative. For example, consider Eq. (8.4), as represented in Eq. (8.6).

$$\begin{aligned} \frac{d^2 P_{t,k}^{\text{EPC}}}{\left(dP_{t,k}^{\text{load}}\right)^2} &= \frac{d^2}{\left(dP_{t,k}^{\text{load}}\right)^2} \int_{P_{t,k}^{\text{load}}}^{P_{\max^{L,k}}} \left(x - P_{t,k}^{\text{load}}\right) f(x) dx \\ &= \frac{d}{dP_{t,k}^{\text{load}}} \left[- \int_{P_{t,k}^{\text{load}}}^{P_{\max^{L,k}}} f(x) dx + P_{t,k}^{\text{load}} f\left(P_{t,k}^{\text{load}}\right) - P_{t,k}^{\text{load}} f\left(P_{t,k}^{\text{load}}\right) \right] \\ &= f\left(P_{t,k}^{\text{load}}\right) \geq 0 \end{aligned} \tag{8.6}$$

The non-negativity of the PDF ensures that $P_{t,k}^{\text{EPC}}$ is a convex function. Consequently, Eqs. (8.4) and (8.5) can be transformed into linearized expressions. Specifically, Eq. (8.4) can be relaxed to Eq. (8.7).

$$P_{t,k}^{\text{EPC}} \geq \int_{P_{t,k}^{\text{Load}}}^{P_{\max^{L,k}}} \left(x - P_{t,k}^{\text{Load}}\right) f(x) dx \tag{8.7}$$

To further linearize Eq. (8.7), the s-block piecewise linear technique is used, as suggested by [32]. Since the objective of this chapter is to optimize operation costs, this relaxation will not affect the ESS charging/discharging strategy. Thus, the segmented linearization shown in Eq. (8.7) can be formulated as follows:

$$
\begin{aligned}
P_{t,k}^{\mathrm{EPC}} &\geq \max_{t\in T,k\in\mathbb{K},s\in\mathbb{S}}\left(\frac{y_{t,k,s+1}^{\mathrm{EPC}}-y_{t,k,s}^{\mathrm{EPC}}}{x_{t,k,s+1}^{\mathrm{EPC}}-x_{t,k,s}^{\mathrm{EPC}}}P_{t,k}^{\mathrm{load}}-x_{t,k,s}^{\mathrm{EPC}}\frac{y_{t,k,s+1}^{\mathrm{EPC}}-y_{t,k,s}^{\mathrm{EPC}}}{x_{t,k,s+1}^{\mathrm{EPC}}-x_{t,k,s}^{\mathrm{EPC}}}+y_{t,k,s}^{\mathrm{EPC}}\right)\\
x_{t,k,s}^{\mathrm{EPC}} &= (s-1)\Delta s\\
y_{t,k,s}^{\mathrm{EPC}} &= \int_{x_{t,k,s}^{\mathrm{EPC}}}^{P_{\max}L}\left(X-x_{t,k,s}^{\mathrm{EPC}}\right)f(X)dX
\end{aligned}
\tag{8.8}
$$

where $\mathbb{K}=\{1,2,\ldots,k,\ldots,K\}$ and $\mathbb{S}=\{1,2,\ldots,s,\ldots,S\}$. S represents the number of linearized segments, while Δs denotes segment length. The intermediate variables $x_{t,k,s}^{\mathrm{EPC}}$ and $y_{t,k,s}^{\mathrm{EPC}}$ are employed, and Fig. 8.2 provides the segment plot of the piecewise linearization technique.

Similarly, the linearized expression of $P_{t,k}^{\mathrm{EPNS}}$ can be derived as:

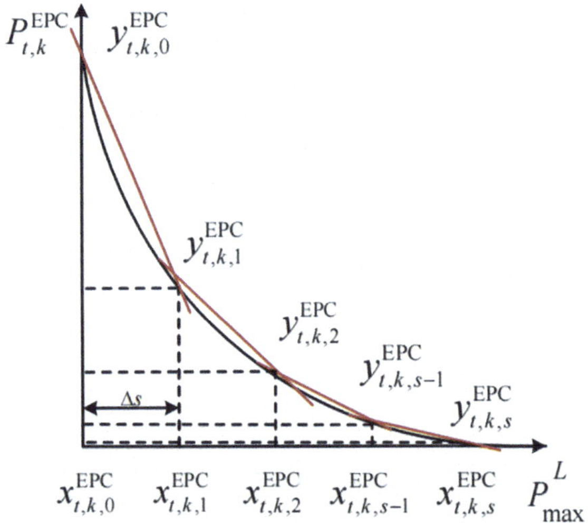

Fig. 8.2 The segment plot of piecewise linearization for $P_{t,k}^{\mathrm{EPC}}$, reprinted from Ref. [35]. Copyright 2024, with permission Journal of Energy Storage

$$P_{t,k}^{\text{EPNS}} \geq \max_{t\in T,k\in\mathbb{K},s\in\mathbb{S}} \left(\frac{y_{t,k,s+1}^{\text{EPNS}} - y_{t,k,s}^{\text{EPNS}}}{x_{t,k,s+1}^{\text{EPNS}} - x_{t,k,s}^{\text{EPNS}}} P_{t,k}^{\text{Load}} - x_{t,k,s}^{\text{EPNS}} \frac{y_{t,k,s+1}^{\text{EPNS}} - y_{t,k,s}^{\text{EPNS}}}{x_{t,k,s+1}^{\text{EPNS}} - x_{t,k,s}^{\text{EPNS}}} + y_{t,k,s}^{\text{EPNS}} \right) \tag{8.9}$$

8.3.2 *Expected Cost Minimization Driven Random Clustering*

8.3.2.1 The Operation Strategy of ESS Considering Load Uncertainty

The substantial variation in load curves across different days makes frequent adjustments to the ESS operation within the industrial park economically impractical. Additionally, daily ESS operation strategies may not be optimal due to potential forecasting errors in load predictions. To address these challenges, this chapter introduces an expected cost minimization-based random clustering method. This approach aims to determine the optimal number of clusters for the load curves, enabling the development of an effective day-ahead operation for ESS tailored to each identified cluster.

The objective is to minimize the total operational cost of the industrial park over a month, encompassing electricity tariffs, demand tariffs, EPNS tariffs, EPC tariffs, and on-grid tariffs. Therefore, the model designed to account for uncertain load conditions can be formulated as:

$$\begin{aligned}
\text{obj} \quad F_0 = \min \sum_{k\in\mathbb{K}} \sum_{g\in G_k} \sum_{t\in T} \Big[& n_k \left(\rho_t P_{t,k}^{\text{buy}} - \rho^{\text{TG}} P_{t,k}^{\text{sell}} \right) \\
& + \rho_t \left(P_{t,g}^{\text{IDB}} + P_{t,k}^{\text{EPC}} \right) \\
& + \rho^{\text{TG}} \left(P_{t,g}^{\text{IDS}} - P_{t,k}^{\text{EPNS}} \right) \Big] + \rho^{\text{DP}} \max_{t\in T, g\in G} \left\{ P_{t,k}^{\text{buy}} + P_{t,g}^{\text{IDB}} \right\}
\end{aligned}$$

$$\begin{aligned}
\text{s.t.} \quad & SOE_{t+1,k} = SOE_{t,k} + \eta^{\text{ch}} P_{t,k}^{\text{ch}} - P_{t,k}^{\text{dch}} / \eta^{\text{dch}} \\
& SOE_{0,k} = SOE_{24,k} \\
& SOC_{\min} \leq SOC_{t,k} \leq SOC_{\max} \\
& 0 \leq P_{t,k}^{\text{ch}} \leq P_{\max} \\
& 0 \leq P_{t,k}^{\text{dch}} \leq P_{\max} \\
& P_{t,k}^{\text{ch}} P_{t,k}^{\text{dch}} = 0 \\
& P_{\min^{L,k}} \leq P_{t,k}^{\text{load}} \leq P_{\max^{L,k}} \\
& P_{t,k}^{\text{net}} = P_{t,k}^{\text{buy}} - P_{t,k}^{\text{sell}} - P_{t,k}^{\text{ch}} + P_{t,k}^{\text{dch}} \\
& P_{t,k}^{\text{buy}} P_{t,k}^{\text{sell}} = 0 \\
& P_{t,g}^{\text{IDB}} = \max \left\{ P_{t,g}^{\text{load}} - P_{t,k}^{\text{load}}, 0 \right\}
\end{aligned}$$

$$P_{t,g}^{\mathrm{IDS}} = \min\left\{P_{t,g}^{\mathrm{load}} - P_{t,k}^{\mathrm{load}}, 0\right\} \tag{8.10}$$

where n_k denotes the number of days associated with the load curve belonging to the set G_k; $P_{t,k}^{\mathrm{buy}}$ and $P_{t,k}^{\mathrm{sell}}$ indicate the purchased power from the upper grid and the sold power to the upper grid at time t with respect to the k th cluster, respectively; $P_{t,g}^{\mathrm{IDB}}$ refers to the power that the industrial park needs to purchase power from the upper grid at time t. It corresponding to that the intra-day load of the industrial park on the g th day is running with the optimal ESS operation with the k th load curve; $P_{t,g}^{\mathrm{IDS}}$ represents the on-grid power at time t, indicating the intra-day load of the industrial park on the g th day corresponding to the optimal ESS operation of the k th load curve. $SOE_{t,k}$ and $SOE_{t+1,k}$ are respectively the ESS's remaining power of at time t and $t + 1$; η^{ch} and η^{dch} denote ESS's charging and discharging efficiencies, respectively; $P_{t,k}^{\mathrm{ch}}$ and $P_{t,k}^{\mathrm{dch}}$ represent ESS's charging and discharging power of at time t when the industrial park runs within the k th cluster load curve; $SOE_{0,k}$ refers to the ESS initial state, and $SOE_{\min}$ and $SOE_{\max}$ represent the minimum and maximum bounds of the SOE. $P_{\max}$ indicates the maximum charging/discharging power. $P_{t,k}^{\mathrm{net}}$ signifies the net load power at time t associated with the k th cluster.

8.3.2.2 The Optimal Clusters Based on Random Clustering

In this chapter, the K-means clustering method is applied to group the monthly load data of the industrial park. The cluster center values, denoted as K, fall within the range $[2, [\sqrt{n}]]$, where n denotes the sampling number and $[\sqrt{n}]$ is an integer indicating the number of clusters. The K-means clustering method requires the optimal number of clusters to be determined manually [33]. The elbow method is commonly used for this purpose, as it identifies the optimal number of clusters based on changes in the slope of the error curve [34].

The traditional elbow method evaluates the sum of squared errors between data points and their respective cluster centers, without considering the economic implications of the system. In contrast, this chapter introduces an enhanced approach by incorporating the actual operational cost of the industrial park. The "elbow point" is developed by analyzing the slope change in the sum of cost error (SCE), which accounts for deviations between the actual load curve and the corresponding optimal load curve for each cluster number. The formula for calculating SCE is:

$$\begin{aligned} SCE = &\sum_{k\in\mathbb{K}}\sum_{g\in G_k}\sum_{t\in T}\left[\rho_t\left(P_{t,g}^{\mathrm{IDB}} + P_{t,g}^{\mathrm{EPC}}\right) + \rho^{\mathrm{TG}}\left(P_{t,g}^{\mathrm{IDS}} - P_{t,g}^{\mathrm{EPNS}}\right)\right] \\ &+ \rho^{\mathrm{DP}} \max_{t\in T, g\in G}\left\{P_{t,k}^{\mathrm{buy}} + P_{t,g}^{\mathrm{IDB}}\right\} \end{aligned} \tag{8.11}$$

where $P_{t,g}^{\mathrm{EPC}}$ and $P_{t,g}^{\mathrm{EPNS}}$ represent EPC and EPNS of the g th day, respectively. As K increases, the members within each cluster become more tightly grouped, leading to a more refined division of samples. Consequently, the SCE value will approach zero.

When the *SCE* decreases very slowly with additional clusters, it indicates that further increases in the number of clusters do not significantly improve clustering effectiveness. Thus, the "elbow point"—the point at which the reduction in SCE becomes marginal—is identified as the optimal clusters number. Subsequently, the chapter employs the K ESS operation curves with respect to this optimal cluster number as the ESS charging/discharging strategy. The detailed process for the clustering is outlined in Algorithm 1.

Algorithm 1 Solution process of optimal cluster

1: Input: K = 2;
2: The ESS operation strategy corresponding to K cluster centers is obtained by Eq. (8.10);
3: The SCE value is calculated by Eq. (8.11);
4: $K = K + 1$;
5: **if** $K \leq \lfloor \sqrt{n} \rfloor$ **then**
6: Enter Step 2
7: **else**
8: The *SCE* values corresponding to $K \in \left[2, \lfloor \sqrt{n} \rfloor\right]$ are plotted;
9: **end if**
10: The optimal cluster number is determined by the elbow method

8.4 Stochastic Optimization with Random Clustering and Dynamic Recognition

In this section, a dynamic recognition technology is presented that leverages several key characteristics, including value range, slope similarity, curve step similarity, and power magnitude similarity to identify the cluster to which an intra-day load curve belongs. This technology enables the determination of the optimal intra-day operation strategy for the ESS.

8.4.1 Value Range Characteristic

The load $L = \left\{L_1, L_2, \ldots, L_j, \ldots, L_N\right\}$ at N consecutive time intervals of the intra-day are randomly sampled, where L_j denotes the load demand at time j. Take the load $Y^{\mathrm{VRC}} = \left\{P_{1,g}^{\mathrm{load}}, P_{2,g}^{\mathrm{load}}, \ldots, P_{j,g}^{\mathrm{load}}, \ldots, P_{N,g}^{\mathrm{load}}\right\}$ of all load curves from G_k for the same time interval, where $P_{j,g}^{\mathrm{load}}$ denotes the load value at time j of day g. If the intra-day load is within k, then the load falls between the minimum and maximum of the cluster. The value range characteristic (VRC) is defined as follows in Eq. (8.12):

$$\min_{g \in G_k}\left\{P_{j,g}^{\text{load}}\right\} \leq L_j \leq \max_{g \in G_k}\left\{P_{j,g}^{\text{load}}\right\} \tag{8.12}$$

Cluster centers that satisfy Eq. (8.12) are retained. If only one cluster center meets this equation, the intra-day load is assigned to the corresponding cluster. If multiple cluster centers satisfy this condition, the slope similarity characteristic (SSC) is applied to these cluster centers that meet VRC.

8.4.2 Slope Similarity Characteristic

Assuming that K_1 cluster centers satisfy Eq. (8.12), the slope with respect to the adjacent moments of the intra-day load is obtained. The slope associated with L_j and L_{j+1} in intra-day load indicates as $A^L = \left\{a_1^L, \ldots, a_j^L, \ldots, a_{N-1}^L\right\}$. The slope matrix combining by adjacent moments of K_1 cluster centers denotes as $Y^{\text{SSC}} = \left\{P_{1,(K_1)}^{\text{load}}, \ldots, P_{j,(K_1)}^{\text{load}}, \ldots, P_{N,(K_1)}^{\text{load}}\right\}_{N \times K_1}$, where $P_{j,(K_1)}^{\text{load}}$ represents the value of K_1 cluster centers at moment j. The slope with respect to $P_{j,(K_1)}^{\text{load}}$ and $P_{j+1,(K_1)}^{\text{load}}$ is $A^{\text{SSC}} = \left\{a_{1,(K_1)}, \ldots, a_{j,(K_1)}, \ldots, a_{N-1,(K_1)}\right\}_{N \times K_1}$.

The direction of the slope t_j^L and $t_{j(K_1)}$ for each load segment is represented as 1, −1, or 0. Specifically, a positive slope between two points is assigned as 1, a negative slope as −1, and a zero slope as 0.

To ascertain whether the intra-day load curve belongs to a specific cluster center, it is compared with each segment t_j^L and $t_{j(K_1)}$ of the K_1 cluster center. Let $H_{j(K_1)}$ denote the similarity between the intra-day load curve and the K_1 cluster center based on the SSC. A similarity value of 1 is assigned if the slope of the intra-day load curve matches the slope of the K_1 cluster center for the corresponding time interval, and 0 otherwise. The number of segments $E_{(K_1)}$ that exhibit the same slope direction as the K_1 cluster center can be computed using Eqs. (8.13) and (8.14):

$$H_{j(K_1)} = \begin{cases} 1, t_j^L = t_{j(K_1)} \\ 0, t_j^L \neq t_{j(K_1)} \end{cases} \tag{8.13}$$

$$E_{(K_1)} = \sum_{j=1}^{N-1} H_{j(K_1)} \tag{8.14}$$

The similarity *SSC* associated with a cluster center and the load curve increases as the value of $E_{(K_1)}$ rises. If the value of $E_{(K_1)}$ meets or exceeds a predetermined threshold $N \times \beta_1$, the cluster center is retained, where β_1 denotes the coefficient for slope similarity. Otherwise, it is excluded. If there is no cluster centers meeting this condition, the input data will be subdivided into a smaller time interval. Then, VRC is applied once more. When only one cluster center meets the revised criteria, the intra-day load curve is assigned to the corresponding cluster center. Conversely, if there still several cluster centers satisfy the VRC condition, the Power Magnitude Similarity Characteristic (PMSC) developed in the following subsection is employed to further differentiate among the cluster centers that meet SSC condition.

8.4.3 Power Magnitude Similarity Characteristic

Assuming that K_2 clusters meet SSC condition, and the load data with respect to K_2 cluster centers denotes as $Y^{\text{PMSC}} = \left\{P^{\text{load}}_{1,(K_2)}, \ldots, P^{\text{load}}_{2,(K_2)}, \ldots, P^{\text{load}}_{j,(K_2)}, \ldots, P^{\text{load}}_{N,(K_2)}\right\}_{N \times K_2}$. Then, calculate the power differences between cluster centers, and their absolute value are represented as $\Delta P = \left\{\Delta P_{1,(K_2)}, \ldots, \Delta P_{j,(K_2)}, \ldots, \Delta P_{N,(K_2)}\right\}_{N \times K_2}$. And the minimum is noted as 1, the rest is noted as 0:

$$Z_{j(K_2)} = \begin{cases} 1, \text{ if } \Delta P_{j(K_2)} = \min\left\{\Delta P_{j(K_2)}\right\} \\ 0, \text{ otherwise} \end{cases} \tag{8.15}$$

Note $P_{(K_2)}$ as the sum of minimum absolute difference:

$$P_{(K_2)} = \sum_{j=1}^{N-1} Z_j^c \tag{8.16}$$

$P_{(K_2)}$ reflects the degree of similarity associated with the cluster center and the intra-day load curve. A higher $P_{(K_2)}$ represents a larger similarity. If the similarity meets the criterion $P_{(K_2)} \geq N \times \beta_2$, the cluster center is retained, where β_2 is the weight coefficient. Otherwise, it is discarded. If there is no cluster center meets this criterion, the load data will be redivided into much smaller time moments. Meanwhile, VRC is reassessed. Should there only one cluster center meet the revised condition, the intra-day load is assigned to the corresponding cluster center. Conversely, if there still several cluster centers meet the criterion, the curve step similarity characteristic (CSSC) is then applied to the cluster centers that satisfy the PMSC.

8.4.4 Curve Step Similarity Characteristic

Assuming that K_3 clusters meet the PMSC criterion, corresponding to load data $Y^{\text{CSSC}} = \left\{P^{\text{load}}_{1,(K_3)}, P^{\text{load}}_{2,(K_3)}, \ldots, P^{\text{load}}_{j,(K_3)}, \ldots, P^{\text{load}}_{N,(K_3)}\right\}_{N\times K_3}$. The distance $l^L = \left\{l^L_1, l^L_2, \ldots, l^L_j, \ldots, l^L_{N-1}\right\}$ between L_j and $L_j + 1$, and the distance $Y^{\text{CSSC}} = \left\{P^{\text{load}}_{1,(K_3)}, P^{\text{load}}_{2,(K_3)}, \ldots, P^{\text{load}}_{j,(K_3)}, \ldots, P^{\text{load}}_{N,(K_3)}\right\}_{N\times K_3}$ between $l_{j,(K_3)}$ and $l_{j+1,(K_3)}$ are calculated. The calculation method with time interval Δt is shown as follows:

$$l^L_j = \sqrt{\Delta t^2 + \left(L_{j+1} - L_j\right)^2} \tag{8.17}$$

$$l^{\text{CSSC}}_j = \sqrt{\Delta t^2 + \left(P^{\text{load}}_{j,(K_3)} - P^{\text{load}}_{j+1,(K_3)}\right)^2} \tag{8.18}$$

The difference between l^L_j and $l_{j,(K_3)}$ is computed using Eq. (8.19). Record the distance corresponding to the smallest value as 1, while other distances are recorded as 0. This process is mathematically represented by Eq. (8.20):

$$\Delta l_{j,(K_3)} = l_{j,(K_3)} - l^L_j \tag{8.19}$$

$$q_{j,(K_3)} = \begin{cases} 1, \text{ if } \Delta l_{j,(K_3)} = \min\left\{\Delta l_{j,(K_3)}\right\} \\ 0, \text{ otherwise} \end{cases} \tag{8.20}$$

The number $Q_{(K_3)}$ corresponding to the smallest difference associated with the curve step size and the daily load among the K_3 cluster centers is determined:

$$Q_{(K_3)} = \sum_{i=1}^{N-1} q_{j,(K_3)} \tag{8.21}$$

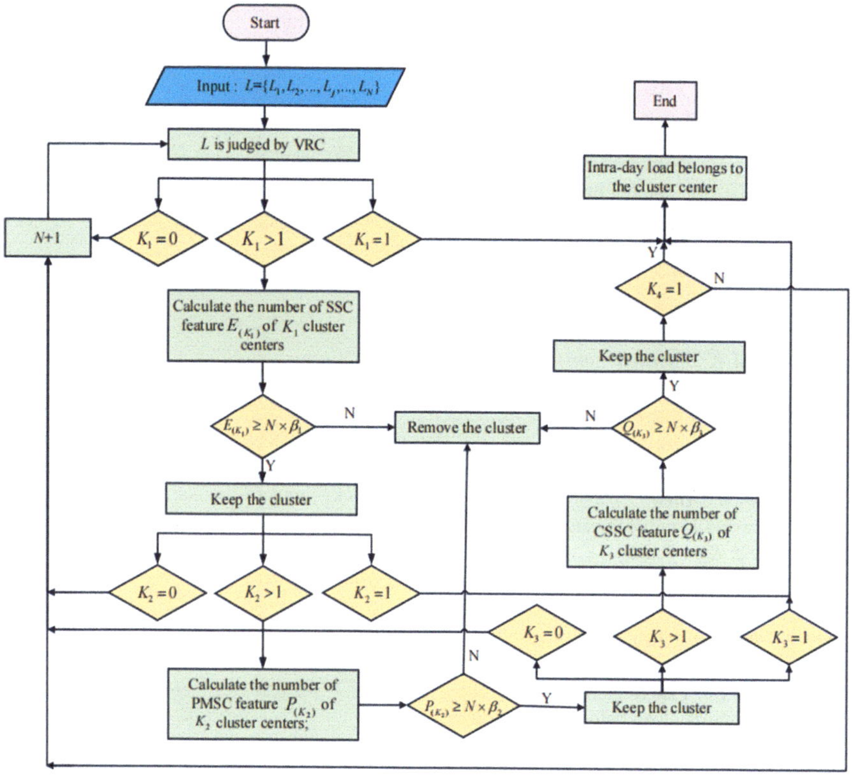

Fig. 8.3 Based on multi-feature load recognition technology flow chart, reprinted from Ref. [35]. Copyright 2024, with permission Journal of Energy Storage

As the value of $Q_{(K_3)}$ increases, the similarity improves. If the cluster center's step size similarity, denoted as $Q_{(K_3)} \geq N \times \beta_3$, aligns with that of the intra-day load, retain the cluster center; otherwise, remove it. β_3 denotes the weight coefficient. If there is no cluster centers meet this criterion, additional time intervals are introduced to the input load, and the aforementioned criterions are re-evaluated. If there is only one cluster center meets the criterion, the intra-day load will be assigned to that cluster center.

Algorithm 2 outlines the process for the proposed dynamic recognition, and the corresponding flowchart is illustrated in Fig. 8.3.

Algorithm 2 VRC, SSC, PMSC, CSSC-based dynamic recognition

Input :$L = L_1, L_2, \ldots, L_j, \ldots, L_N$
L is judged by VRC;
if $K_1 = 1$ **then**
  Intra-day load belongs to the cluster center;
else if $K_1 > 1$ **then**
  Calculate the number of SSC $E_{(K1)}$ of K_1 cluster centers;
  if $E_{(K_1)} \geq N \times \beta_1$ **then**
  Keep the cluster;
   end if
   if $K_2 = 1$ **then**
   Intra-day load belongs to the cluster center;
   else if $K_2 > 1$ **then**
   Calculate the number of PMSC $P_{(K2)}$ of K_2 cluster centers;
   if $P_{(K_2)} \geq N \times \beta_2$ **then**
   Keep the cluster;
   end if
   if $K_3 = 1$ **then**
   Intra-day load belongs to the cluster center;
   else if $K_3 > 1$ **then**
    Calculate the number of CSSC $Q_{(K3)}$ of K_3 cluster centers;
    if $P_{(K_3)} \geq N \times \beta_3$ **then**
     Keep the cluster;
    end if
    if $K_4 = 1$ **then**
     Intra-day load belongs to the cluster center;
    else
      Update input data, and perform VRC;
    else
      Update input data, and perform VRC;
    else
      Update input data, and perform VRC;
    end if
   end if
  end if
end if

8.4.5 Normalized Multi-characteristic Weight Calculation

The calculation process for the weight coefficients of slope similarity, power similarity, and step similarity involves three steps:

Step 1: Calculate the E, P, and Q with respect to one-month load data and K cluster centers by Eqs. (8.14), (8.16) and (8.21), respectively. Then, normalize these data.

Step 2: Calculate mean values of normalized E, P, and Q.

Step 3: Determine the weight coefficients β_1, β_2, and β_3 for each characteristic. These coefficients are computed as the percentage of each characteristic's mean value relative to the total mean value of all characteristics.

8.5 Case Study

8.5.1 *Parameter Setting*

One month's electricity load from an industrial park in Shandong province, China is employed. K ranges from 2 to 6 [2, 6]. The ESS is 60 kW/215 kWh, with a charging/discharging efficiency of 0.9, and a SOC range of [0.2, 1]. The industrial park consists of time-of-use tariff and demand tariff with 38 CNY/kWh. The time-of-use price is set as follows:

- Shoulder tariff: 1.1533 CNY/kWh, applicable from 10:00–11:00 and 16:00–21:00.
- Peak tariff: 0.9857 CNY/kWh, applicable from 09:00–10:00, 15:00–16:00, and 21:00–22:00.
- Off-peak tariff: 0.7027 CNY/kWh, applicable from 07:00–09:00, 11:00–12:00, and 14:00–15:00.
- Flat tariff: 0.4197 CNY/kWh, applicable from 00:00–07:00 and 12:00–14:00.

8.5.2 *Optimal Cluster Number and ESS Operation Strategy*

Figure 8.4 depicts the variation in *SCE* with different cluster numbers. As observed, *SCE* decreases with an increase in clusters number. The steepest change in SCE slope occurs at $K = 3$. According to the elbow method, $K = 3$.

Figures 8.5(a–c) illustrate different cluster center for $k = \{1, 2, 3\}$. The ESS operation for $k = 3$ is depicted in Fig. 8.6. The hours from 22:00 to 24:00 and from 00:00 to 07:00 denote the valley periods. ESS charging time illustrated in Fig. 8.6 is primarily concentrated. In addition, the hours from 08:00 to 10:00 and from 18:00 to 21:00 indicate the peak periods. Therefore, ESS discharging is primarily concentrated in these periods.

8.5.3 *ESS Operation with and Without Considering Load Uncertainty*

The simulation results with and without considering load uncertainty are presented in Table 8.1. As shown, load uncertainty affects the park system's costs differently

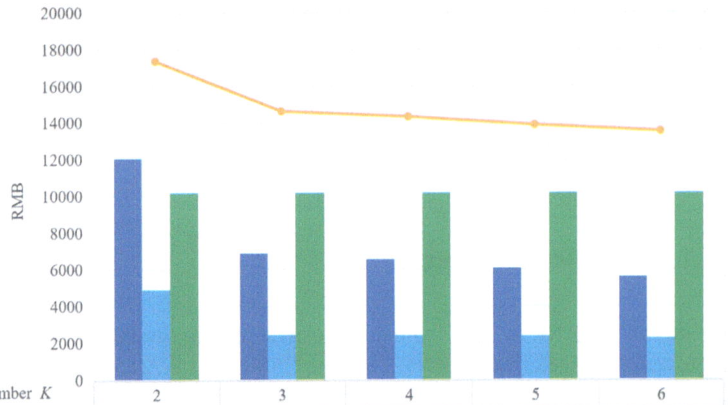

Cluster number K	2	3	4	5	6
Electricity traiff deviation	12071.44	6896.2	6575.97	6067.38	5618.72
On-grid traiff deviation	4895.13	2439.22	2421.71	2361.96	2250.46
Demand traiff deviation	10219.81	10219.81	10219.81	10219.81	10219.81
SCE	17396.12	14676.79	14374.07	13925.23	13588.07

Fig. 8.4 SCE values under different cluster numbers of load curves

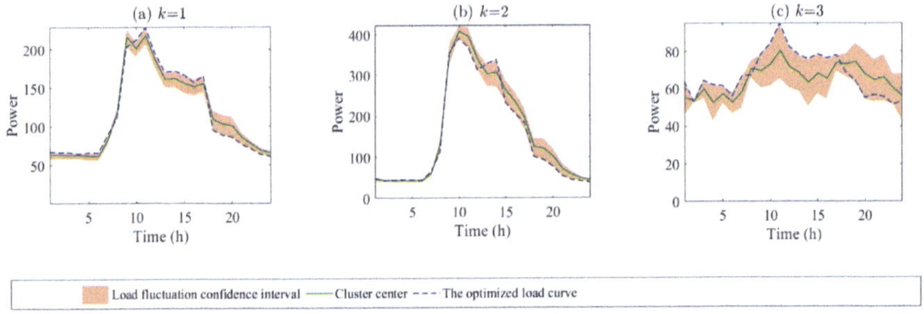

Fig. 8.5 The cluster center, the load curves of the cluster center, the fluctuation range of the load, and the optimal load curve at $K = 3$

even when the K value remains constant. Specifically, load uncertainty has minimal impact on C_{demand}, a relatively minor effect on C_{ND} and C_{sell}, and a more significant impact on C_{buy} and C_{PD}. Notably, the operation of the ESS, optimized for each load curve in consideration of uncertainty, results in a reduction of C_{buy} and an increase of C_{PD} compared to the ESS operation under deterministic loads. Overall, factoring in load uncertainty reduces the operating cost.

The detailed optimization results with and without considering load uncertainty are provided in Table 8.2. The results reveal that accounting for load uncertainty results in a load reduction of 61.76 kW. This indicates that load uncertainty brings highly risk for the industrial park.

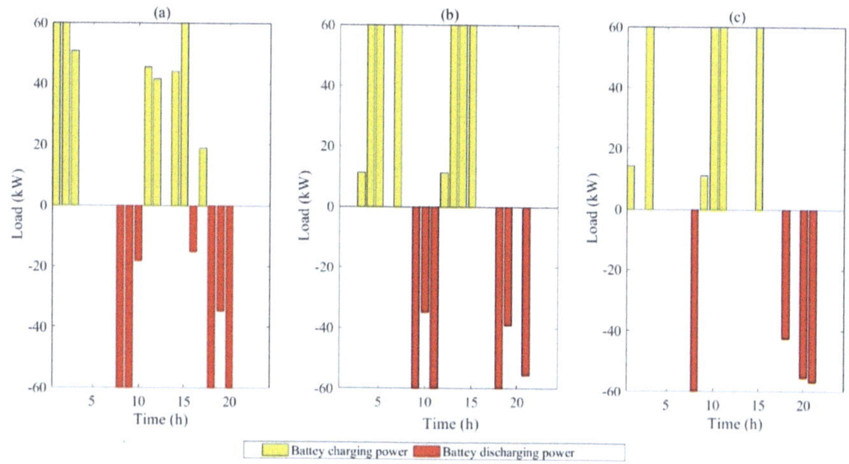

Fig. 8.6 The ESS operation curve of the park system at $K = 3$

Table 8.1 The operating cost of industrial park in one month under different cluster numbers of power load

K	Scenario	Electricity tariff C_{buy} (CNY)	On-grid tariff C_{sell} (CNY)	Demand tariff C_{demand} (CNY)	Positive deviation tariff CPD (CNY)	Negative deviation tariff CND (CNY)	Total tariff C_{cost} (CNY)
2	Deterministic	27,081.54	1759.13	10,219.81	9318.64	4567.67	40,293.19
	Uncertain	22,064.86	904.41	10,268.25	13,503.1	5264.79	39,667.01
3	Deterministic	29,870.19	3074.01	10,219.81	5077.16	2405.03	39,688.12
	Uncertain	26,251.61	2537.57	10,232.34	7430.82	2508.11	38,869.09
4	Deterministic	29,792.99	3074.01	10,219.81	4678.06	2258.61	39,358.24
	Uncertain	26,271.71	2537.57	10,228.57	7129.61	2517.99	38,574.33
5	Deterministic	29,821.55	3087.93	10,219.81	4226.96	2075.7	39,104.69
	Uncertain	26,491.6	2516.87	10,228.57	6593.87	2473.88	38,323.29

8.5.4 Dynamic Recognition for Load Curve

This subsection tries to verify the effectiveness of dynamic recognition through four cases. In Case 1, load curves $\{P^{\text{load}}_{t,g(G_1)}, P^{\text{load}}_{t,g(G_2)}, P^{\text{load}}_{t,g(G_3)}\}$ are randomly selected from each cluster's load curve within a fixed time period. In Case 2, the load data from random time periods is selected. Case 3 involves curves $\{P^{\text{load}}_{t,g'(G_1)}, P^{\text{load}}_{t,g'(G_2)}, P^{\text{load}}_{t,g'(G_3)}\}$ between the minimum and maximum load of each cluster center with a fixed period. In Case 4, the load data from random time periods is selected while keeping load curves $\{P^{\text{load}}_{t,g'(G_1)}, P^{\text{load}}_{t,g'(G_2)}, P^{\text{load}}_{t,g'(G_3)}\}$ fixed. Table 8.4 provides the recognition results for each characteristic quantity in the four cases (Table 8.3).

Table 8.2 Comparison of results with and without considering uncertain load, reprinted from ref. [35]. Copyright 2024, with permission Journal of Energy Storage

Model	Deterministic load					Uncertain load						
Hour	Cluster center (kW)			P^{load} (kW)		P^{EPC} (kW)				$P_{t,k}^{\text{EPNS}}$ (kW)		
	$k = 1$	$k = 2$	$k = 3$	$k = 1$	$k = 2$	$k = 3$	$k = 1$	$k = 2$	$k = 3$	$k = 1$	$k = 2$	$k = 3$
1	63.19	46.75	55.18	68.04	50.20	64.08	0	0	0	0.23	0.16	0.81
2	63.42	43.55	53.55	66.99	44.17	53.33	0	0.03	0.01	0.15	0.05	0
3	62.70	43.07	59.75	66.84	44.92	64.61	0	0.01	0	0.17	0.09	0.22
4	62.75	43.05	52.40	65.02	46.12	62.34	0	0	0	0.09	0.14	1.06
5	61.80	43.06	57.08	66.76	46.05	61.88	0	0	0	0.25	0.13	0.23
6	61.50	42.95	52.58	66.42	44.92	56.16	0	0.01	0.01	0.25	0.10	0.23
7	83.56	60.58	58.32	90.23	66.48	66.99	0	0	0	0.33	0.37	0.73
8	119.32	131.63	70.84	112.12	113.87	67.64	0.39	2.13	0.17	0	0	0
9	216.45	352.05	69.59	204.59	357.85	78.65	0.59	0.33	0	0	0.53	0.66
10	201.56	406.81	73.17	212.69	391.13	84.69	0	0.54	0	0.38	0	1.03
11	218.44	394.79	80.39	229.69	370.35	95.15	0	1.35	0	0.36	0	1.53
12	185.48	338.16	71.67	195.14	313.33	82.80	0	1.62	0	0.31	0	0.98
13	161.08	302.06	68.64	171.16	329.92	78.49	0	0	0	0.39	1.59	0.80
14	162.02	306.51	63.31	172.58	340.50	76.20	0	0	0	0.43	2.33	1.48
15	155.34	262.36	68.09	167.28	233.61	78.79	0	2.81	0	0.57	0	0.95
16	151.24	234.63	65.52	157.13	207.38	76.73	0.04	2.82	0	0.29	0	1.08
17	155.33	196.19	73.92	166.23	184.99	78.34	0	1.68	0	0.48	0.21	0.15
18	109.06	124.94	73.07	95.91	102.48	69.10	1.43	3.56	0.26	0	0	0.01
19	102.96	121.21	74.41	89.47	94.77	64.53	1.59	5.14	1.04	0	0	0

(continued)

Table 8.2 (continued)

Model	Deterministic load					Uncertain load						
Hour	Cluster center (kW)			P^{load} (kW)		P^{EPC} (kW)				$P_{t,k}^{\text{EPNS}}$ (kW)		
	$k = 1$	$k = 2$	$k = 3$	$k = 1$	$k = 2$	$k = 3$	$k = 1$	$k = 2$	$k = 3$	$k = 1$	$k = 2$	$k = 3$
20	100.51	101.47	67.69	86.96	78.46	55.37	1.65	4.64	1.79	0	0	0.01
21	85.75	70.99	64.79	77.54	55.61	56.89	0.37	2.97	1.07	0	0	0.03
22	76.99	56.92	66.22	71.34	45.81	55.93	0.28	1.93	1.49	0	0	0.02
23	68.73	48.34	60.11	64.08	42.58	51.79	0.34	0.61	0.86	0	0	0
24	65.47	44.21	55.85	61.93	39.75	53.45	0.17	0.40	0.74	0	0	0.24
Total	2794.65	3816.26	1556.17	2826.13	3645.23	1633.96	7.24	32.61	7.43	4.68	5.66	12.25
Total		8167.08			8105.32			47.28			22.59	

Table 8.3 PV local consumption power with and without considering uncertain load, reprinted from ref. [35]. Copyright 2024, with permission Journal of Energy Storage

	Load deterministic (kW)			Load uncertain (kW)		
	$k = 1$	$k = 2$	$k = 3$	$k = 1$	$k = 2$	$k = 3$
10	123.43	123.43	123.43	123.43	123.43	89.76
11	194.45	194.45	73.17	194.45	194.45	144.69
12	275.15	275.15	140.39	275.15	275.15	155.15
13	236.65	236.65	131.67	236.65	236.65	82.80
14	166.13	319.95	68.64	171.16	319.95	78.49
15	222.02	282.07	63.31	216.72	282.07	76.20
16	155.34	229.14	85.37	227.28	229.14	138.79
17	141.94	141.94	65.52	141.93	141.94	76.73
18	184.99	184.99	73.92	184.9 9	184.99	78.34
Total	1700.10	1987.76	825.42	1771.75	1987.76	920.95
Total		4513.28			4680.46	

The simulation of $P^{\text{load}}_{t,g(G_1)}$ in Case 1 is discussed as follows. VRC screening is firstly applied, retaining cluster centers $k = 1$ and $k = 2$, while eliminating $k = 3$. Subsequently, calculate SSC for $k = 1$ and $k = 2$. Since $E^1 \geq N \times \beta_1$ and $E^2 \geq N \times \beta_1$, $k = 1$ and $k = 2$ are retained. Then, calculate PMSC for $k = 1$ and $k = 2$. As $P^1 \geq N \times \beta_2$ and $P^2 \geq N \times \beta_2$, $k = 1$ and $k = 2$ are retained. Finally, calculate CSSC for $k = 1$ and $k = 2$. Since $k = 1$ and $k = 2$ are not satisfied, the load curve $P^{\text{load}}_{t,g(G_1)}$ belongs to the first category. Therefore, ESS operation for the day is determined to be $k = 1$.

Figure 8.7(a–c) present the load curves $P^{\text{load}}_{t,g(G_1)}$, $P^{\text{load}}_{t,g(G_2)}$ of Case 1 and Case 2. Figure 8.7(d–f) display the load $P^{\text{load}}_{t,g(G_1)}$, $P^{\text{load}}_{t,g(G_2)}$ of Case 3 and Case 4, along with the clustering centers and ESS charging/discharging curves recognized by the same approach. The results on these cases validate the effectiveness and accuracy of the proposed dynamic recognition for reliably identifying random load curves at random time periods.

Table 8.4 The recognition results of each characteristic based on dynamic recognition under Cases 1–4, reprinted from ref. [35]. Copyright 2024, with permission Journal of Energy Storage

$P^{load}_{t,g(G_1)}$						$P^{load}_{t,g(G_2)}$						$P^{load}_{t,g(G_3)}$						
Case 1			Case2			Case 1			Case 2			Case 1			Case 2			
	$k=1$	$k=2$	$k=3$	$k=1$	$k=2$	$k=3$	$k=1$	$k=2$	$k=3$	$k=1$	$k=2$	$k=3$	$k=1$	$k=2$	$k=3$	$k=1$	$k=2$	$k=3$
VRC	5	5	0	5	1	4	5	5	0	4	5	2	2	1	5	5	5	5
SSC	4	4	–	–	–	–	4	4	–	–	–	–	–	–	–	2	2	4
PMSC	2	3	–	–	–	–	1	4	–	–	–	–	–	–	–	–	–	–
CSSC	3	1	–	–	–	–	–	–	–	–	–	–	–	–	–	–	–	–
The recognized																		
Cluster center	✓			✓				✓			✓				✓			✓
$P^{load}_{t,g'(G_1)}$						$P^{load}_{t,g'(G_2)}$						$P^{load}_{t,g'(G_3)}$						
Case 3			Case 4			Case 3			Case 4			Case 3			Case 4			
	$k=1$	$k=2$	$k=3$	$k=1$	$k=2$	$k=3$	$k=1$	$k=2$	$k=3$	$k=1$	$k=2$	$k=3$	$k=1$	$k=2$	$k=3$	$k=1$	$k=2$	$k=3$
VRC	5	5	2	5	1	2	3	5	2	5	5	1	2	0	5	5	5	5
SSC	3	3	–	–	–	–	–	–	–	3	3	–	–	–	–	3	3	3
PMSC	3	2	–	–	–	–	–	–	–	1	3	–	–	–	–	0	1	4
CSSC	4	1	–	–	–	–	–	–	–	–	–	–	–	–	–	–	–	–
The recognized																		
Cluster center	✓			✓				✓			✓				✓			✓

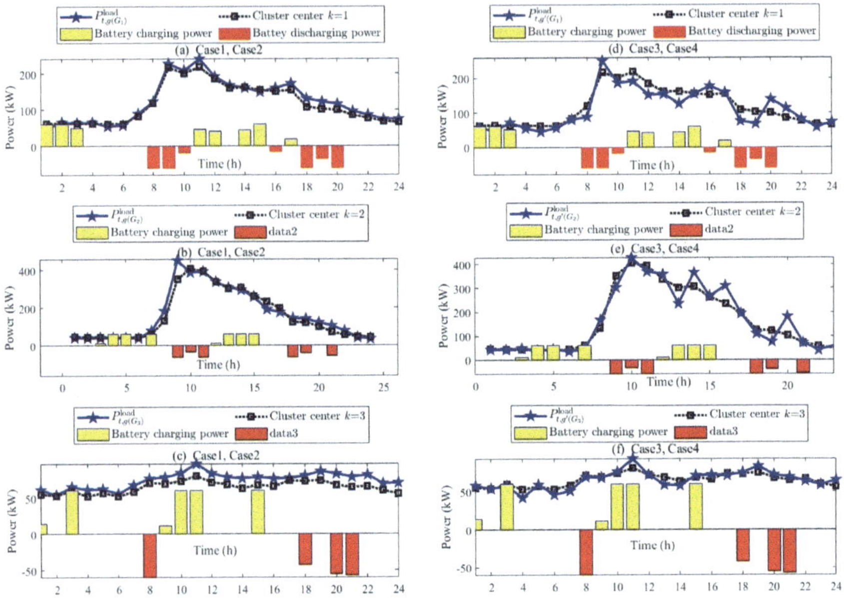

Fig. 8.7 The optimized load curves and ESS operation strategies under Cases 1–4

8.6 Conclusion

This chapter introduces a random clustering and dynamic recognition-based ESS operation for industrial parks. The random clustering, aimed at minimizing expected costs, considers linear risk measure and load costs to derive the optimal number of clusters and the corresponding ESS charging/discharging strategy. The dynamic recognition, encompassing four characteristics, identifies the cluster of the intra-day load curve and generates the ESS operation strategy. The main findings are:

(1) The random clustering accounts for the uncertainty risk of park load and costs, ensuring optimal operation for the park. This approach surpasses the constraints of current clustering methods, which often neglect practical operational limitations such as load uncertainty and operational expenses.

(2) By employing relaxation and piecewise linearization techniques, the nonlinear EPNS and EPC models are transformed into mixed-integer linear problems, significantly enhancing computational efficiency. □

(3) The dynamic recognition adeptly pinpoints clusters within intra-day load curves by extracting various characteristics, including value range, slope similarity, power size similarity, and curve step similarity. This method effectively tackles the challenges arising from the highly volatile and intermittent nature of power loads.

References

1. Fang J, Xu Q, Tang R, Xia Y, Ding Y, Fang L (2021) Research on demand management of hybrid energy storage system in industrial park based on variational mode decomposition and wigner–ville distribution. J Energy Storage 42:103073
2. Feng L, Zhang X, Li C, Li X, Li B, Ding J et al (2022) Optimization analysis of energy storage application based on electricity price arbitrage and ancillary services. J Energy Storage 55:105508
3. Zhang X, Qin CC, Xu Y, Li W, Zhou X, Li R et al (2021) Integration of small-scale compressed air energy storage with wind generation for flexible household power supply. J Energy Storage 37:102430
4. Thokar RA, Gupta N, Niazi K, Swarnkar A, Meena NK (2021) Multiobjective nested optimization frame-work for simultaneous integration of multiple photovoltaic and battery energy storage systems in distribution networks. J Energy Storage 35:102263
5. Schmidt O, Hawkes A, Gambhir A, Staffell I (2017) The future cost of electrical energy storage based on experience rates. Nat Energy 2(8):1–8
6. Zhang M, Chen J, Yang Z, Peng K, Zhao Y, Zhang X (2021) Stochastic day-ahead scheduling of irrigation system integrated agricultural microgrid with pumped storage and uncertain wind power. Energy 237:121638
7. Zhang Z, Zhou K, Yang S (2023) Optimal selection of energy storage system sharing schemes in industrial parks considering battery degradation. J Energy Storage 57:106215
8. Wen G, Yu X, Liu ZW, Yu W (2017) Adaptive consensus-based robust strategy for economic dispatch of smart grids subject to communication uncertainties. IEEE Trans Industr Inf 14(6):2484–2496
9. Ren H, Wu Q, Gao W, Zhou W (2016) Optimal operation of a grid-connected hybrid pv/fuel cell/battery energy system for residential applications. Energy 113:702–712
10. Liu J, Chen J, Yan G, Chen W, Xu B (2023) Clustering and dynamic recognition based auto-reservoir neural network: a wait-and-see approach for short-term park power load forecasting. iScience 26:107456
11. Yang J, Su C (2021) Robust optimization of microgrid based on renewable distributed power generation and load demand uncertainty. Energy 223:120043
12. Chen J, Qi B, Rong Z, Peng K, Zhao Y, Zhang X (2021) Multi-energy coordinated microgrid scheduling with integrated demand response for flexibility improvement. Energy 217:119387
13. Chen J, Wu Q, Zhang L, Wu P (2017) Multi-objective mean–variance–skewness model for nonconvex and stochastic optimal power flow considering wind power and load uncertainties. Eur J Oper Res 263(2):719–732
14. Jiao P, Chen J, Cai X, Wang L, Zhao Y, Zhang X et al (2021) Joint active and reactive for allocation of renewable energy and energy storage under uncertain coupling. Appl Energy 302:117582
15. Zou Y, Xu Y, Zhang C (2023) A risk-averse adaptive stochastic optimization method for transactive energy management of a multi-energy microgrid. IEEE Trans Sustain Energy 14(6):1–12
16. Fazlalipour P, Ehsan M, Mohammadi-Ivatloo B (2019) Risk-aware stochastic bidding strategy of renewable micro-grids in day-ahead and real-time markets. Energy 171:689–700
17. Zhang Y, Wang J, Zeng B, Hu Z (2017) Chance-constrained two-stage unit commitment under uncertain load and wind power output using bilinear benders decomposition. IEEE Trans Power Syst 32(5):3637–3647
18. Qu Z, Chen J, Peng K, Zhao Y, Rong Z, Zhang M (2021) Enhancing stochastic multi-microgrid operational flexibility with mobile energy storage system and power transaction. Sustain Cities Soc 71:102962
19. Wright A, Firth S (2007) The nature of domestic electricity-loads and effects of time averaging on statistics and on-site generation calculations. Appl Energy 84(4):389–403
20. Guo Y, Tong L, Wu W, Zhang B, Sun H (2017) Coordinated multi-area economic dispatch via critical region projection. IEEE Trans Power Syst 32(5):3736–3746

21. Williams JL, Fisher JW, Willsky AS (2007) Approximate dynamic programming for communication-constrained sensor network management. IEEE Trans Sig Process 55(8):4300–4311
22. Ojaghi P, Rahmani M (2017) Lmi-based robust predictive load frequency control for power systems with communication delays. IEEE Trans Power Syst 32(5):4091–4100
23. Flath C, Nicolay D, Conte T, van Dinther C, Filipova-Neumann L (2012) Cluster analysis of smart metering data: an implementation in practice. Bus Inf Syst Eng 4:31–39
24. Zhao Q, Li H, Wang X, Pu T, Wang J (2019) Analysis of users electricity consumption behavior based on ensemble clustering. Glob Energy Interconnection 2(6):479–488
25. Kaur R, Gabrijelčič D (2022) Behavior segmentation of electricity consumption patterns: a cluster analytical approach. Knowl-Based Syst 251:109236
26. Rezaei N, Pezhmani Y, Mohammadiani RP (2023) Optimal stochastic self-scheduling of a water-energy virtual power plant considering data clustering and multiple storage systems. J Energy Storage 65:107366
27. Parvizimosaed M, Farmani F, Monsef H, Rahimi-Kian A (2017) A multi-stage smart energy management system under multiple uncertainties: a data mining approach. Renew Energy 102:178–189
28. Allahvirdizadeh Y, Galvani S, Shayanfar H (2021) Data clustering based probabilistic optimal scheduling of an energy hub considering risk-averse. Int J Electr Power Energy Syst 128:106774
29. Rawa M, Al-Turki Y, Sedraoui K, Dadfar S, Khaki M (2023) Optimal operation and stochastic scheduling of renewable energy of a microgrid with optimal sizing of battery energy storage considering cost reduction. J Energy Storage
30. Zhang Z, Chow MY (2012) Convergence analysis of the incremental cost consensus algorithm under different communication network topologies in a smart grid. IEEE Trans Power Syst 27(4):1761–1768
31. Wen G, Yu X, Liu ZW, Yu W (2017) Adaptive consensus-based robust strategy for economic dispatch of smart grids subject to communication uncertainties. IEEE Trans Ind Inf (99):1
32. Zimmerman RD, Murillo-Sánchez CE, Thomas RJ (2010) Matpower: steady-state operations, planning, and analysis tools for power systems research and education. IEEE Trans Power Syst 26(1):12–9
33. Sinaga KP, Yang MS (2020) Unsupervised k-means clustering algorithm. IEEE Access 8:80716–80727
34. Liu F, Deng Y (2020) Determine the number of unknown targets in open world based on elbow method. IEEE Trans Fuzzy Syst 29(5):986–995
35. Yan GJ, Chen JJ, Liu JY et al (2023) Random clustering and dynamic recognition-based operation strategy for energy storage system in industrial park[J]. J Energy Storage 73:109192

Chapter 9
Mobile Energy Storage System Optimization with Peer-to-Peer for Resilience Improvement

9.1 Introduction

The safe and stable supply of electricity is a crucial driver of contemporary economic and social development. Reducing or even avoiding power system failures is essential for maintaining economic growth and social stability [1–3]. As the terminal segment of the power grid, the distribution system plays a vital role. Sudden failures in this system can directly impact the power supply reliability for connected users and potentially cause widespread blackouts [4–6]. Consequently, enhancing the resilience of the distribution system against sudden failures, such as extreme weather events and operational errors, has garnered significant attention from scholars both domestically and internationally [7–11].

The energy storage system (ESS) is a key technological approach to enhancing the resilience of distribution systems. Reference [12] examined the reduction of load loss during natural disasters through the strategic configuration of fixed ESS. The study demonstrated that optimal placement and charging/discharging power settings for fixed ESS can significantly mitigate load loss during disasters. However, traditional fixed ESS located outside fault areas cannot support users within those areas during faults. Reference [13] explored the optimal configuration of mobile energy storage (MES) capacity in scenarios involving multiple line faults, finding that MES can reduce load loss by 73.2%. MES, as a backup resource, can create dynamic microgrids post-failure, providing continuous power to critical loads in fault zones. Reference [14] proposed a resilience enhancement scheme using active MES aggregation, showing effective fault loss reduction through emergency MES scheduling. Additionally, Reference [15] investigated MES real-time scheduling strategies to improve resilience under various fault conditions. The findings indicate that MES can enhance system resilience; however, previous studies have not accounted for the impact of user distance and MES mobility speed on its spatial and temporal performance.

J. Chen and Y. Li, *Renewable Power System Optimization*, Smart Energy Systems,
https://doi.org/10.1007/978-981-97-8132-4_9

In the prevalent distributed power operation model of spontaneous self-use, MES typically interacts with users within its immediate vicinity and seldom shares energy with users located elsewhere in the distribution system. The peer-to-peer (P2P) transaction model, however, offers a platform for energy interaction and sharing between MES and users throughout the distribution network [16, 17]. Integrating P2P transactions with MES can address the issue of fault load loss by leveraging the P2P platform to facilitate power sharing across different locations. Additionally, this approach can reduce overall system operational costs, presenting a viable method for enhancing the resilience of distribution systems [18]. Thus, optimizing the spatial and temporal characteristics of MES through effective P2P power sharing is crucial for improving system resilience. Furthermore, establishing a fair cost-sharing mechanism for users participating in MES power sharing is essential.

This chapter proposes a strategy to enhance distribution system resilience through MES sharing within the P2P transaction framework. It explores how power sharing between MES and users in different locations can mitigate fault-related losses. Additionally, a cost-sharing mechanism based on proportional distribution is introduced, calculating the operation and maintenance costs that each user should bear according to their share of energy from MES.

9.2 P2P Trading and MES Sharing Model

9.2.1 Mathematical Model of P2P Transaction

In P2P transaction, distributed power supplies can share energy with users located in other areas [19]. There are two primary P2P trading frameworks: the system-centric model, which aims to maximize social welfare, and the user-centric model, which focuses on maximizing individual user benefits [20, 21]. Given that a unified scheduling decision is crucial for ensuring continuous power supply to critical loads following a line fault, this chapter adopts a system-centric P2P trading framework. This approach integrates MES, distributed power sources, and other resources into the centralized scheduling of the distribution system, enabling energy sharing under the P2P model. The objective is to minimize fault-related load loss and reduce overall operational costs.

In the system-centric P2P model, the goal of maximizing social welfare translates to minimizing the operating costs of the distribution system. These costs encompass the power generation expenses of each user's distributed power supply and the costs associated with purchasing electricity from the main grid [21].

(1) The objective function of P2P trading mode:

$$\min \sum_{t \in T} \left(\sum_{b \in N} C_t^w P_{t,b}^{\text{grid}} + \sum_{g \in G_b} C_g P_{t,b,g} \right) \tag{9.1}$$

where T represents the set of scheduling periods t; n denotes the set of distribution system users b; C_g is the set of distributed generation units g within the distribution system; G_b denotes the power generation cost for distributed generation unit g; C_t^w is the grid electricity price during time period t; $P_{t,b,g}$ represents the active power output of distributed generation unit g at user b during period t; and $P_{t,b}^{\text{grid}}$ is the active power purchased by user b from the grid in period t.

(2) P2P transaction constraints:

$$\begin{cases} P_{t,b} = \sum\limits_{\omega \in \Omega_t \& \omega_s = b} \rho_{\omega,t}^p, D_{t,b}^p = \sum\limits_{\omega \in \Omega_t \& \omega_H = b} \rho_{\omega,t}^p \\ Q_{t,b} = \sum\limits_{\omega \in \Omega_t \& \omega_s = b} \rho_{\omega,t}^q, D_{t,b}^q = \sum\limits_{\omega \in \Omega_t \& \omega_H = b} \rho_{\omega,t}^q \end{cases}, \quad \forall b \in N \tag{9.2}$$

where $P_{t,b}$ and $Q_{t,b}$ are the active power and reactive power transmitted by user b through P2P transaction in time period t; $D_{t,b}^p$ and $D_{t,b}^q$ denote the active power and reactive power obtained by user b through P2P transaction in period t, respectively. Ω_t indicates the set of electricity sharing ω in time period t; $\omega_s = b$ and $\omega_H = b$ represent the transmission party and the acquisition party of the power sharing ω, respectively; $\rho_{\omega,t}^p$ and $\rho_{\omega,t}^q$ are the active power and reactive power corresponding to the period t power sharing ω, respectively.

The P2P transaction aims to optimize the system's power structure, minimize network losses, and reduce operational costs by leveraging the locational advantages and power generation costs of each distributed power supply. By applying the objective function and constraints of the P2P transactions to the distribution system, it is possible to determine the optimal power sharing between distributed power supplies and users, as well as the amount of power each user needs to purchase from the grid, all while achieving the minimum operating cost for the distribution system.

9.2.2 Mathematical Model of MES Power Sharing

As a distributed energy storage system, the spatial and temporal characteristics of MES distinguish it from other distributed power sources. The ability of MES to rapidly engage in fault repair and ensure power supply within fault areas of the distribution system relies heavily on its spatial and temporal transfer characteristics [15]. Additionally, optimizing MES charging and discharging under price mechanisms can reduce the overall operating costs of the distribution system. This chapter addresses these aspects by constructing a mathematical model that describes the power sharing of MES within the distribution system, incorporating constraints related to operating states, spatial–temporal transfer, and charging-discharging operations.

(1) MES running status:

Unlike the MES charging and discharging models discussed in References [22, 23], the MES power sharing model presented in this chapter encompasses not only the charging and discharging states but also the moving state:

$$\sum_{b\in N}\sum_{t\in T}\left(S^{c}_{t,b,k}+S^{d}_{t,b,k}+\sum_{b'\in N}S^{m}_{t,b,b',k}\right)=1,\quad \forall k\in K \tag{9.3}$$

where K denotes the set of MES; $S^{c}_{t,b,k}$, $S^{d}_{t,b,k}$, $S^{m}_{t,b,b',k}$ represent $\{0, 1\}$ binary integer decision variables; $S^{c}_{t,b,k}$ and $S^{d}_{t,b,k}$ indicate the charging signal and discharge signal of the kth MES at the time period t at the user b; $S^{m}_{t,b,b',k}$ signifies the mobile signal from user b to user b' of the kth MES in time period t.

Based on the operational state constraints of MES outlined above, it is evident that each MES can only maintain a single operational state at any given time for each user.

(2) MES space–time transfer constraints:

In the space–time transfer process, the driving time of MES is influenced by factors such as driving distance, speed, and various uncertainties like traffic congestion and road conditions. Additionally, upon reaching the destination, MES often requires some installation time to facilitate electrical energy interaction with the user. To address these factors, this chapter first calculates the shortest path matrix D between users in the distribution system using Dijkstra's algorithm [24–26]. Subsequently, the time–space transfer process of MES is modeled considering the average driving speed, traffic congestion at different times, and installation time. The MES time–space transfer model is as follows:

$$\varepsilon_{t,b,b'}=\begin{cases}\frac{D_{b,b'}}{v^{\mathrm{mes}}}+T^{\mathrm{ini}}+T^{\mathrm{d}}_{t}, & b\neq b'\in N\\ 0, & b=b'\in N\end{cases} \tag{9.4}$$

$$\begin{cases}L_{t,b',k}=\sum\limits_{b\in N}\left(S^{\mathrm{m}}_{t-1,b,b',k}\zeta_{t,b,b'}\right)+S_{t,b',k}L_{t-1,b',k}\\ \sum\limits_{b'\in N}\left(S^{\mathrm{m}}_{t-1,b,b',k}\zeta_{t,b,b'}\right)+S_{t,b,k}=1\\ \sum\limits_{b\in N}L_{t,b,k}=1\ \forall b\neq b'\in N,\forall k\in K\end{cases} \tag{9.5}$$

where $\varepsilon_{t,b,b'}$ is the time required for MES to travel from user b to user b' during the period t; $D_{b,b'}$ is the shortest distance between user b and user b'; v^{mes} is the average speed of MES; T^{ini} is the time required to install MES; T^{d}_{t} is the traffic congestion delay caused by the traffic congestion degree of time period t; both $S_{t,b,k}$ and $L_{t,b,k}$

are {0–1} binary integer decision variables, $L_{t,b,k}$ is the real-time location of MES. If the kth MES in t period is located at user b, then $L_{t,b,k} = 1$, otherwise, $L_{t,b,k} = 0$, $S_{t,b,k}$ is an auxiliary variable describing the real-time location; $\zeta_{t,b,b'}$ is the auxiliary variable of whether MES can be driven from user b to user b' in Δt. If $\varepsilon_{t,b,b'} \leq \Delta t$, then $\zeta_{t,b,b'} = 1$, that is, MES can be driven from user b to user b' in Δt, otherwise, $\zeta_{t,b,b'} = 0$.

In the MES space–time transfer constraints, Eq. (9.4) calculates the shortest travel time of MES between two users, accounting for the traffic congestion level. Meanwhile, Eq. (9.5) enables the real-time positioning of MES by appropriately setting the binary integer decision variables.

(3) MES charge and discharge constraints:

MES needs to meet its maximum power constraint, state of charge constraint and power balance constraint during charging and discharging.

$$S_{t,k}^{\text{soe}} = S_{t-1,k}^{\text{soe}} + \sum_{b \in N} (P_{t,b,k}^{\text{ch}} - P_{t,b,k}^{\text{dis}}), \quad \forall k \in K \tag{9.6}$$

$$S_{k,\min}^{\text{soc}} \leq S_{t,k}^{\text{soc}} = \frac{S_{t,k}^{\text{soe}}}{E_k} \leq S_{k,\max}^{\text{soc}}, \quad \forall k \in K \tag{9.7}$$

$$0 \leq P_{t,b,k}^{\text{ch}} \leq P_{\max} S_{t,b,k}^{c} L_{t,b,k}, \quad \forall k \in K, \quad \forall b \in N \tag{9.8}$$

$$0 \leq Q_{t,b,k}^{\text{ch}} \leq Q_{\max} S_{t,b,k}^{c} L_{t,b,k}, \quad \forall k \in K, \quad \forall b \in N \tag{9.9}$$

$$0 \leq P_{t,b,k}^{\text{dis}} \leq P_{\max} S_{t,b,k}^{d} L_{t,b,k}, \quad \forall k \in K, \quad \forall b \in N \tag{9.10}$$

$$0 \leq Q_{t,b,k}^{\text{dis}} \leq Q_{\max} S_{t,b,k}^{d} L_{t,b,k}, \quad \forall k \in K, \quad \forall b \in N \tag{9.11}$$

$$S_{0,k}^{\text{soe}} = S_{24,k}^{\text{soe}}, \quad \forall k \in K \tag{9.12}$$

where $S_{t,k}^{\text{soe}}$ and $S_{t,k}^{\text{soc}}$ are the residual power and state of charge of the kth MES in the period t, respectively; $P_{t,b,k}^{\text{ch}}$ and $Q_{t,b,k}^{\text{ch}}$ are the active power and reactive power of the kth MES when charging at the user b during the period t; $P_{t,b,k}^{\text{dis}}$ and $Q_{t,b,k}^{\text{dis}}$ represent the active power and reactive power of the kth MES when the period t discharges at user b, respectively. $P_{\max}$ and $Q_{\max}$ denote the maximum active power and reactive power of MES charging and discharging, respectively. $S_{k,\max}^{\text{soc}}$ and $S_{k,\min}^{\text{soc}}$ indicate the maximum and minimum values of the state of charge of the kth MES, respectively. E_k is the rated capacity of the kth MES.

Equation (9.6) shows that the electric energy of MES in t period is related to the electric energy in $t-1$ period, and the charging and discharging amount in t period. Equation (9.7) is the state of charge constraint of MES in t period. Equations (9.8)–(9.11) are the active and reactive power limits of MES charging and discharging. When $S^{c}_{t,b,k}$ and $L_{t,b,k}$ are both 1, it means that the kth MES in t period is charged at user b. When $S^{\mathrm{d}}_{t,b,k}$ and $L_{t,b,k}$ are both 1, it means that the kth MES in t period discharges at user b. Equation (9.12) indicates that MES needs to meet the power balance constraint in one day.

9.3 Improving Resilience Based on MES Sharing Under P2P

In the P2P transaction mode, MES facilitates power sharing with each user. Given that the available electric energy may not satisfy the load demands of all users simultaneously, it is essential to strategically allocate the limited power resources within the distribution system. This allocation must be optimized to enhance the overall resilience of the distribution system.

(1) Objective function:

When a sudden fault occurs in the distribution system, minimizing the operational costs associated with the fault is a critical measure of the system's resilience. This chapter, therefore, proposes a strategy to minimize the total operational cost following a fault. This includes costs related to power generation, load loss, emergency response involving MES, and the costs associated with the movement of MES.

$$\min \sum_{t\in T}\sum_{b\in N}\sum_{g\in G_b} C_g P_{t,b,g} + \sum_{t\in T}\sum_{b\in N} C^{w}_{t} P^{\mathrm{grid}}_{t,b} + C^{\mathrm{voll}}_{b}\sum_{t\in T}\sum_{b\in N^{\mathrm{fault}}} (P^{d}_{t,b} - p_{t,b})\Delta t + C^{d}\sum_{t\in T}\sum_{b\in N}\sum_{b'\in N}\sum_{k\in K} S^{m}_{t,b,b',k} \tag{9.13}$$

where $P^{d}_{t,b}$ is the active power demand of user b in time period t; C^{d} denotes the moving cost coefficient of MES unit time interval; Δt indicates the time interval of emergency dispatching of distribution system; C^{voll}_{b} represents the loss of load cost coefficient of distribution system after fault.

In the objective function, the first term represents the generation cost of distributed power supplies during the period t, while the second term accounts for the cost of purchasing electricity from the grid. Following a fault, the distribution system is categorized into two zones: the normal area, which remains connected to the power grid, and the fault area, which is disconnected. Users in the fault area rely on power sharing with MES and distributed power supplies to meet their load demands. If the power available to these users from MES and distributed sources is insufficient to cover their actual load requirements, the resulting shortfall represents the fault loss of load. Thus, the third term reflects the cost associated with this load loss. The fourth term covers the costs related to the MES driving process.

(2) Power sharing constraints:

In the distribution system, both MES and distributed generation participate in power sharing with users through P2P transactions. The constraints governing this power sharing are as follows:

$$\begin{cases} D_{t,b}^{p} = \sum\limits_{\omega \in \Omega_{m,t} \& \omega_H = b} \rho_{\omega,t}^{p} + \sum\limits_{\omega \in \Omega_{d,t} \& \omega_H = b} \rho_{\omega,t}^{p} \\ D_{t,b}^{q} = \sum\limits_{\omega \in \Omega_{m,t} \& \omega_H = b} \rho_{\omega,t}^{q} + \sum\limits_{\omega \in \Omega_{d,t} \& \omega_H = b} \rho_{\omega,t}^{q} \end{cases}, \quad \forall b \in N \tag{9.14}$$

$$\sum_{k \in K} P_{t,b,k}^{\text{dis}} = \sum_{\omega \in \Omega_{m,t} \& \omega_H = b} \rho_{\omega,t}^{p}, \quad \forall b \in N \tag{9.15}$$

$$\sum_{g \in G_b} P_{t,b,g} = \sum_{\omega \in \Omega_{d,t} \& \omega_s = b} \rho_{\omega,t}^{p} + \sum_{k \in K} P_{t,b,k}^{\text{ch}}, \quad \forall b \in N \tag{9.16}$$

where $D_{t,b}^{p}$ and $D_{t,b}^{q}$ represent the active power and reactive power obtained by user b through power sharing in time period t; $\omega \in \Omega_{m,t}$, $\omega_H = b$ denote the set of user b in the energy sharing set of MES in t period; $\omega \in \Omega_{d,t}$, $\omega_H = b$ indicate the set of users b in the energy sharing set of distributed power supply in t period.

(3) Power flow constraints:

After a disconnection fault occurs in the distribution system, the normal area remains electrically connected to the power grid, while the fault area loses its connection. This disconnection may result in an insufficient power supply in the fault area to meet the load demands of all users. Consequently, the power flow constraint of the distribution system is expressed as follows:

$$\begin{cases} f_{t,l}^{\text{p}} \le \partial_l S_l \\ f_{t,l}^{\text{q}} \ge -\partial_l S_l \end{cases}, \quad \forall l \in L \tag{9.17}$$

$$0 = \sum_{k \in K} \left(P_{t,b,k}^{\text{ch}} - P_{t,b,k}^{\text{dis}}\right) - \sum_{l | r(l) = b} f_{t,l}^{q} + p_{t,b} + f_{t,l|s(l)=b}^{p} - P_{t,b}^{\text{grid}} - P_{t,b,g}, \quad \forall b \in N \tag{9.18}$$

$$0 = \sum_{k \in K} \left(Q_{t,b,k}^{\text{ch}} - Q_{t,b,k}^{\text{dis}}\right) - \sum_{l | r(l) = b} f_{t,l}^{q} + q_{t,b} + f_{t,l|s(l)=b}^{q} - Q_{t,b}^{\text{grid}} - Q_{t,b,g}, \quad \forall b \in N \tag{9.19}$$

$$\begin{cases} p_{t,b} = P^d_{t,b}, q_{t,b} = Q^d_{t,b} \ \forall b \in N^{\text{norm}} \\ p_{t,b} = D^p_{t,b}, q_{t,b} = D^q_{t,b} \ \forall b \in N^{\text{fault}} \end{cases} \tag{9.20}$$

where L represents the set of transmission line l; N^{norm} and N^{fault} denote the set of users in the normal area and the fault area, respectively. $f^p_{t,l}$ and $f^q_{t,l}$ are the active power and reactive power transmitted by line l in time period t respectively; ∂_l is a binary integer decision variable. If line l fails, ∂_l is 0; otherwise, ∂_l is 1. S_l is the upper limit of transmission power of line l; $s(l)$ and $r(l)$ are the power outflow end and the inflow end of line l, respectively. $f^p_{t,l|s(l)=b}$ and $f^q_{t,l|s(l)=b}$ are the active power and reactive power transmitted by the line l flowing into the user b, respectively; $f^{\text{p}}_{t,l|r(l)=b}$ and $f^{\text{q}}_{t,l|r(l)=b}$ are the active power and reactive power transmitted by the line l flowing out of user b, respectively; $p_{t,b}$ and $q_{t,b}$ are the active power and reactive power obtained by user b in time period t, respectively.

Equations (9.17)–(9.19) are the power flow constraints of the normal area when MES realizes power sharing in P2P trading. Equation (9.20) denotes the electric energy obtained by each user in t period. Because the power demand of users in the normal area can always be met, the actual power obtained by users in this area is their own demand, while the fault area can only obtain power through P2P transactions, so the power obtained by users in the fault area is the power obtained through P2P transactions.

(4) Distributed power output constraints:

In the process of power sharing between distributed generation and users, it is necessary to meet its own maximum output constraints:

$$\begin{cases} 0 \le P_{t,b,g} \le P_{b,g,\max} \\ 0 \le Q_{t,b,g} \le Q_{b,g,\max} \end{cases}, \forall g \in G_b, \forall b \in N \tag{9.21}$$

where $P_{b,g,\max}$ and $Q_{b,g,\max}$ are the maximum active power and reactive power output of the distributed generation g at the user b, respectively.

The mathematical model of the distribution system resilience improvement strategy proposed in this chapter is composed of (9.1)–(9.21). By solving the mathematical model, the time–space characteristics of MES charging and discharging, the power sharing scheme between MES and each user and each user can be obtained.

9.4 MES Operation and Maintenance Cost

In this chapter, MES units are managed by large power users, and each user can obtain electricity by sharing energy with MES. Consequently, each user is responsible for a proportionate share of the operation and maintenance costs associated with the MES. Following established practices in power system cost allocation, particularly the proportional allocation mechanism as discussed in References [27, 28], we apply this principle to MES cost-sharing.

Reference [27] demonstrates that proportional distribution is an effective method for sharing transmission costs among users, based on their actual usage of transmission lines. Building on this approach, this chapter proposes a cost-sharing mechanism for MES operation and maintenance using the proportional distribution method. This mechanism allocates costs to each user based on their share of the total MES power sharing. Specifically, it determines each user's contribution to the operation and maintenance costs by calculating their proportion of the total MES power shared.

MES operation and maintenance costs:

$$c = \frac{|h|}{100} \sum_{t \in T} \sum_{b \in N} \sum_{k \in K} C^P (P^{\mathrm{ch}}_{t,b,k} + P^{\mathrm{dis}}_{t,b,k}) \tag{9.22}$$

where c is the operation and maintenance cost generated during the operation of MES; both C^P and $|h|$ are MES cost coefficients.

According to Eq. (9.12), MES meets the power balance constraint in one day, so the charging amount of MES in one day is equal to the discharging amount. According to Eq. (9.15), Eq. (9.22) can be expressed as:

$$c = \frac{|h|}{100} \sum_{t \in T} \sum_{b \in N} 2C^P \left(\sum_{\omega \in \Omega_{m,t} \& \omega_H = b} \rho^P_{\omega,t} \right) \tag{9.23}$$

Once the total operation and maintenance cost of MES is determined, it is essential to calculate the cost allocation among the users participating in MES power sharing. The cost allocation formula is given by:

$$c_b = c \frac{\sum_{t \in T} \sum_{\omega \in \Omega_{m,t} \& \omega_H = b} \rho^P_{\omega,t}}{\sum_{t \in T} \sum_{b' \in N} \sum_{\omega \in \Omega_{m,t} \& \omega_H = b'} \rho^P_{\omega,t}}, \forall b \in N \tag{9.24}$$

where c_b is the MES operation and maintenance cost sharing amount of user b.

By applying this cost allocation mechanism, the operation and maintenance expenses incurred by MES can be proportionally distributed among users according to their share of power exchanged with MES within the distribution system.

9.5 Case Study

In this chapter, we conduct simulation analyses for the 15-node, IEEE 33-node, and IEEE 69-node radial distribution systems. The topology, line parameters, and other data for the 15-node radial distribution system are referenced from the literature [21]. The system structure is depicted in Fig. 9.1, and detailed parameters are provided in

Table 9.1. Notably, the 24-h load profiles of the distribution systems are calculated based on load percentages. Simulations and analyses are performed using Julia, with the Gurobi solver employed to solve the problems with a precision of 0.1%.

This chapter begins with the simulation and analysis of the 15-node radial distribution system. In this system, users 2, 4, and 13 are designated as critical loads, representing a hospital, a biological laboratory, and a government facility, respectively. The transmission line has a unit resistance of 0.17 Ω/km [29]. Large power users, specifically user 2 and user 13, are equipped with a Mobile Energy Storage (MES) system, each with a capacity of 3.5 MWh and an initial state of charge of

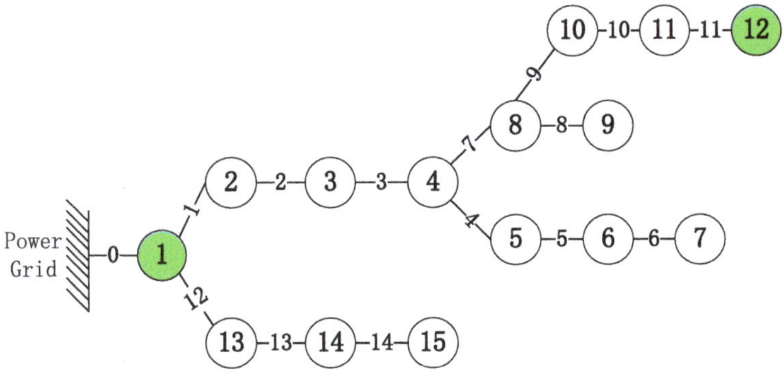

Fig. 9.1 15-node radial power distribution system

Table 9.1 Parameters of 15-node radial power distribution system

Node	P^d (MW)	Q^d (MVar)	P_{max} (MW/h)	Q_{max} (MW/h)	C^{voll} ($/MWh)	Cost ($/MWh)
1	0.0000	0.0000	0.2	0.2	300	40
2	0.7381	0.1725	0	0	600	0
3	0.0000	0.0000	0	0	300	0
4	0.0467	0.0195	0	0	600	0
5	0.0402	0.0100	0	0	300	0
6	0.0677	0.0170	0	0	300	0
7	0.0509	0.0128	0	0	300	0
8	0.0509	0.0044	0	0	300	0
9	0.0546	0.0137	0	0	300	0
10	0.0532	0.0330	0	0	300	0
11	0.0505	0.0151	0	0	300	0
12	0.0307	0.0077	0.6	0.6	300	10
13	0.5784	0.1201	0	0	600	0
14	0.0033	0.0019	0	0	300	0
15	0.0521	0.0193	0	0	300	0

60%. Traffic congestion data for different time periods is sourced from [29], while detailed MES parameter settings are referenced from [30–32].

The chapter focuses on the impact of common line faults caused by emergencies on the distribution system and investigates strategies to enhance resilience in response to various line faults. To evaluate the effectiveness of these resilience improvement strategies, the chapter examines several typical fault conditions, analyzing the spatial–temporal characteristics and power sharing capabilities of MES. It then assesses how different strategies affect the distribution system's resilience.

(1) Line 1 fault:

Figure 9.2a illustrates the real-time charge and discharge amounts of the MES following a fault on line 1, with the histogram representing these quantities and the line chart depicting the MES's state of charge. Figure 9.2b shows the MES's real-time location. Due to the lower cost of distributed power generation at user 12, both MES units are directed there for charging. Once charging is complete, the MES units proceed to user 2 to provide power to users in the fault area. As the distributed power supply in the fault area fails to meet the total load demand, it is necessary for the MES units to supply additional power to other users in the fault area to mitigate load loss.

(2) Line 4 fault:

Figure 9.3a displays the real-time charge and discharge amounts of the MES following a fault on line 4, with the histogram showing these amounts and the line chart depicting the MES's state of charge. Figure 9.3b illustrates the real-time positions of the MES units. Given the low cost of distributed power generation at user 12, when line 4 fails in the distribution system, the MES units first travel to user 12 for charging. Subsequently, they alternate in supplying power to the fault area. Due to the

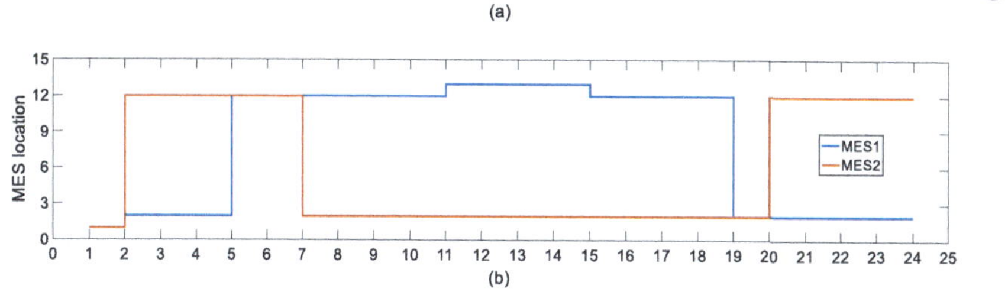

Fig. 9.2 Real-time state of MES after line 1 fault

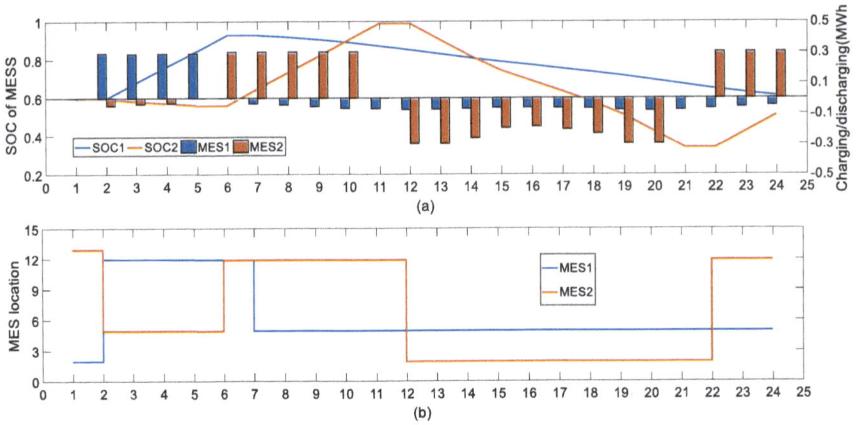

Fig. 9.3 Real-time state of MES after line 1 fault

relatively small power demand in the fault area, while one MES is providing power there, the other MES heads to the normal area to engage in peak-valley arbitrage. This coordinated operation between the MES units helps to minimize fault load loss and reduce the overall operational cost of the distribution system.

(3) Line 12 fault:

Figure 9.4a presents the real-time charge and discharge amounts of the MES units following a fault on line 12, with the histogram depicting these amounts and the line chart illustrating the state of charge of the MES units. Figure 9.4b shows the real-time positions of the MES units. Upon the fault occurrence on line 12, MES2 located at user 13 supplies power to the affected area. Due to the low cost of distributed power generation at user 12, MES1 from user 2 travels there to charge. After MES2 has completed its discharge, MES1 proceeds to the fault area to provide power, ensuring continuous supply to key loads.

In P2P transaction, MES can be centrally managed within the distribution system. By sharing electrical energy with users, this mode effectively reduces fault load loss and enhances the resilience of the distribution system. The following table presents the load losses in the distribution system under various line faults:

The data presented in Table 9.2 indicates that the resilience improvement model proposed for various line fault conditions not only ensures a continuous power supply to key loads but also reduces the load loss rate for general loads to under 5%. Thus, the model introduced in this chapter effectively enhances the resilience of the distribution system, particularly for key loads.

Additionally, the operational costs of the distribution system vary significantly across different fault scenarios. Table 9.3 compares these costs for various faults.

From Table 9.3, it is evident that the load loss cost for the distribution system following a line 4 fault is the lowest. This is attributable to the relatively small load demand in the fault area, which can be adequately supported solely by the MES.

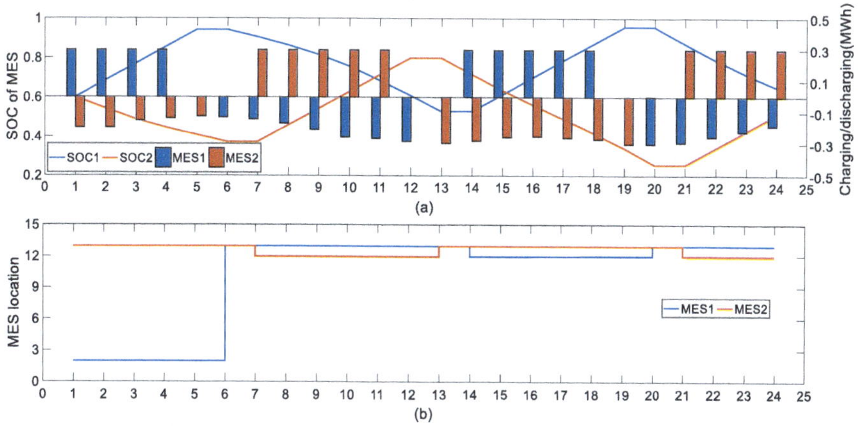

Fig. 9.4 Real-time state of MES after line 1 fault

Table 9.2 Loss of load when different lines are disconnected

Fault line	Load type	Load demand (MW)	Real load (MW)	Loss of load (MW)
1	Critical load	11.958	11.958	0
	General load	3.984	3.817	0.167
4	Critical load	11.958	11.958	0
	General load	3.984	3.871	0.113
12	Critical load	11.958	11.958	0
	General load	3.984	3.825	0.159

Table 9.3 Cost comparison under different failure scenarios

Fault line	Line 1	Line 4	Line 12
Operation cost ($)	356.0	424.9	377.7
Loss of load cost ($)	50.1	33.8	47.8
MES cost ($)	40.0	30.0	30.0
Total cost ($)	446.0	488.6	414.9

Conversely, when line 1 or line 12 fails, the fault area has a higher load demand and includes key loads. During power sharing, MES prioritizes supplying users with key loads, leading to increased load loss among general users. Consequently, the load loss costs for failures of line 1 or line 12 are higher compared to a line 4 fault.

This analysis confirms the effectiveness of the proposed resilience improvement strategy for the 15-node radial distribution system. To further validate the general applicability of this strategy, simulations and analyses of various line faults in IEEE 33-node and IEEE 69-node distribution systems have been conducted. The results of load loss and cost comparisons for different fault scenarios in these systems are presented in Tables 9.4 and 9.5, respectively.

In the IEEE 33-node distribution system, users 4, 7, 8, and 30 are identified as critical loads. Specifically, user 4 represents the government, users 7 and 8 are biological laboratories, and user 30 is a hospital. In this system, MES units are owned by large power users, namely users 8 and 32. Similarly, in the IEEE 69-node distribution system, users 7, 11, 12, 21, 48, and 49 are designated as critical loads, with user 7 being a government facility, users 11 and 12 as hospitals, and users 21, 48, and 49 as biological laboratories. MES units in this system are owned by large power users, specifically users 12 and 48.

According to Tables 9.4 and 9.5, the resilience improvement strategy outlined in this chapter effectively ensures a continuous power supply to critical loads under various line fault scenarios in both the IEEE 33-node and IEEE 69-node systems. Additionally, it successfully reduces the load loss rate for general loads to less than 10%. These results demonstrate the universal applicability of the proposed strategy.

In P2P transaction, power sharing within the distribution system varies significantly depending on the fault line and the time of the fault. This chapter presents the power sharing dynamics at $t = 15$ h post-fault for different line faults. Figures 9.5, 9.6, and 9.7 illustrate the power sharing for line faults 1, 4, and 12, respectively. In these figures, DG1 and DG2 represent distributed power sources located at users 1 and 12.

To evaluate the impact of MES power sharing on the resilience of the distribution system within the P2P transaction framework, this chapter compares the operating costs and fault load losses under three distinct fault conditions across different strategies. The operating costs for these strategies are depicted in Fig. 9.8.

Table 9.4 Loss of load under different fault scenarios in IEEE 33-node and IEEE 69-node systems

Load type	Test system	IEEE 33-node			IEEE 69-node		
	Fault line	6	22	25	10	35	46
General load (MWh)	Load demand	21.717	21.717	21.717	27.569	27.569	27.569
	Loss of load	1.652	1.425	0.896	0.2643	0.1555	0.7517
Critical load (MWh)	Load demand	5.221	5.221	5.221	6.585	6.585	6.585
	Loss of load	0.0	0.0	0.0	0.0	0.0	0.0

Table 9.5 Costs of different failure scenarios in IEEE 33-node and IEEE 69-node systems

Test system	IEEE 33-node			IEEE 69-node		
Fault line	6	22	25	10	35	46
Operation cost ($)	2148.34	2114.74	2198.15	2279.08	2177.35	2256.93
Loss of load cost ($)	495.51	427.41	268.82	79.29	46.67	225.52
MES cost ($)	30.0	50.0	30.0	30.0	15.0	20.0
Total cost ($)	2673.85	2592.15	2496.97	2388.37	2239.02	2502.45

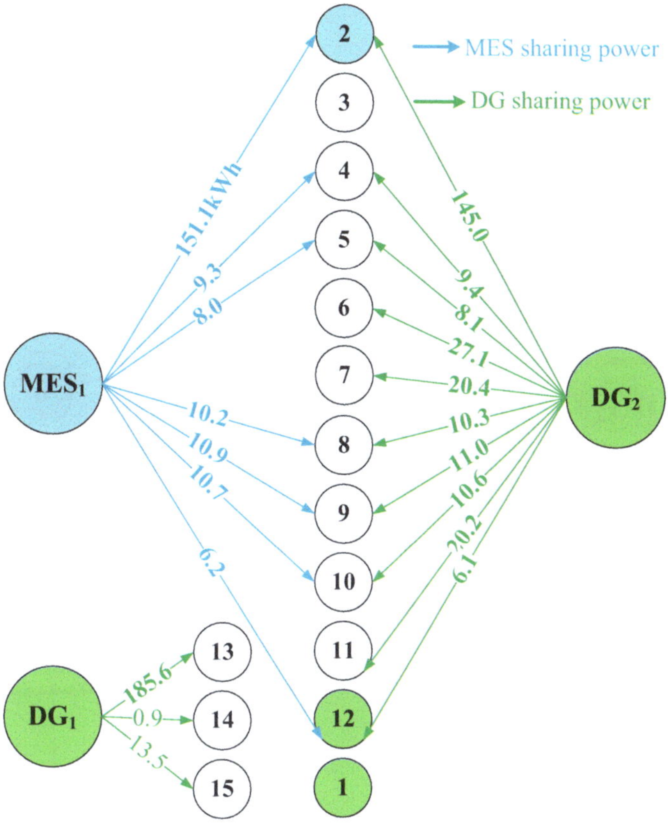

Fig. 9.5 Power sharing of 15 h after line 1 fault

Figure 9.8a, b illustrate the operational and load loss costs of the distribution system under various strategies. In these figures, "P2P + MES" denotes the resilience improvement strategy introduced in this chapter, while "MES" represents a strategy focusing solely on MES without P2P integration. The P2P transaction mode addresses the limitation of users with distributed power supplies being restricted to self-consumption, enabling them to share power with other users. MES leverages its ability to dynamically adjust the spatial and temporal distribution of energy, effectively mitigating power supply shortages in fault areas. Compared to other operational strategies, the approach proposed in this chapter demonstrates significant benefits in reducing both load loss and operational costs of the distribution system.

Following the analysis of resilience improvements under the three fault conditions, the proportional distribution method is employed to allocate the operation and maintenance costs of MES power sharing among users in the distribution system. The cost-sharing mechanism for each fault condition is detailed in Table 9.6.

When line 1 fails, the cost allocation is significantly higher for user 2 due to its critical load and substantial demand. In contrast, when line 4 fails, the load demand

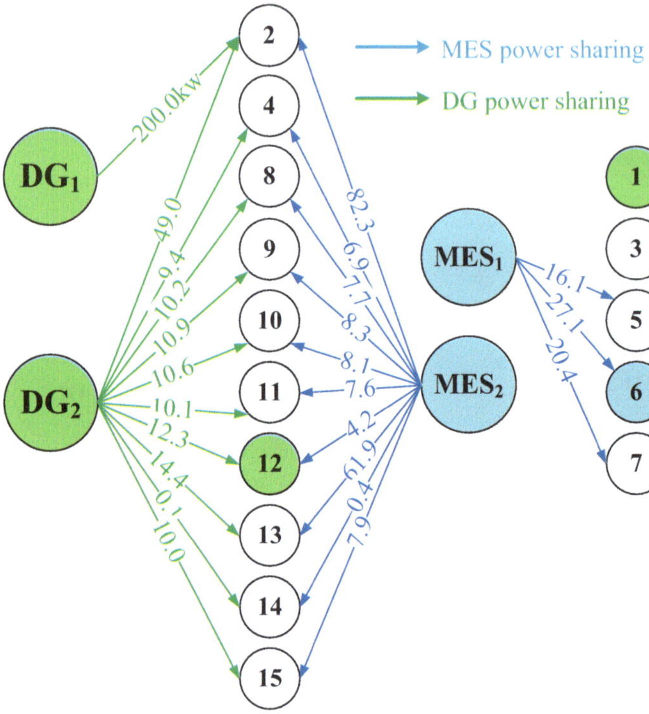

Fig. 9.6 Power sharing of 15 h after line 4 fault

in the fault area is relatively low. One MES can fulfill this demand, while the other MES operates in the normal area, focusing on charging and discharging to minimize the system's operational costs. Consequently, although users in both the fault area and the normal area participate in MES power sharing, all distribution system users, except for users 1 and 3, incur some cost.

Following a line 12 failure, the fault area encompasses users 13, 14, and 15. The cost allocation for user 13 is notably high due to its critical load and significant demand. However, users 2 through 12 in the normal area are not involved in MES power sharing and thus incur no cost.

The discussed scenarios involve single-line faults. To address the possibility of simultaneous multi-line faults, this chapter evaluates the effectiveness of the proposed resilience improvement strategy. The optimal MES charging and discharging strategy, as well as power supply details for various load types, are illustrated in Fig. 9.9 and Table 9.7, respectively. The results demonstrate that even under multiple line faults, the distribution system can maintain continuous power supply to critical loads through MES, distributed power supplies, and power sharing among users within the P2P transaction framework. Additionally, the load loss rate for general loads can be reduced to 19%. Hence, the proposed model is effective for handling multi-line fault scenarios.

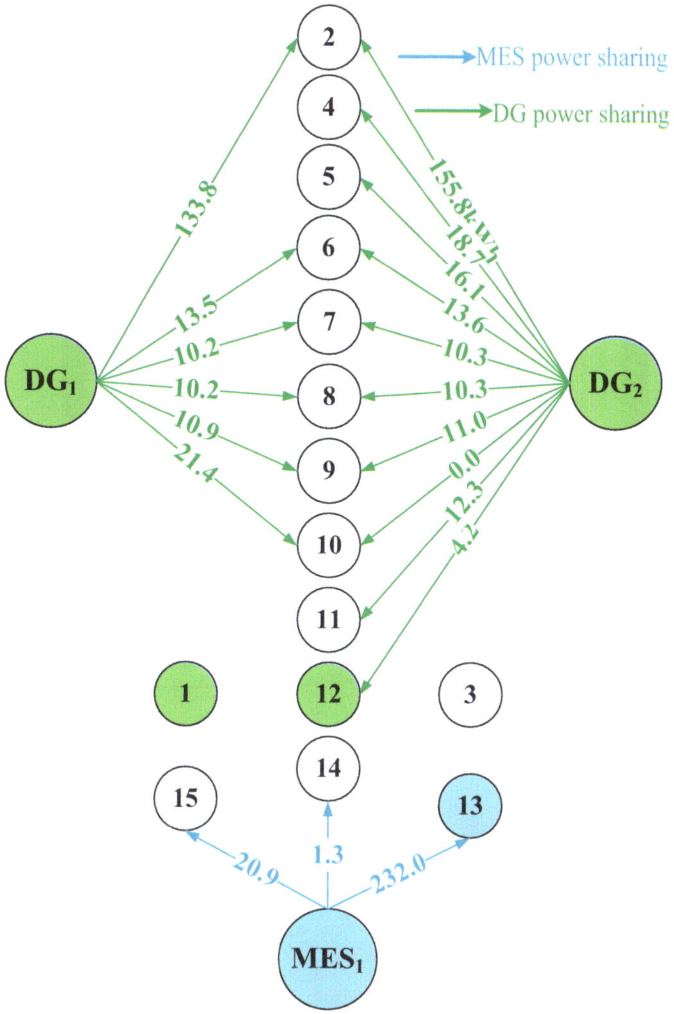

Fig. 9.7 Power sharing of 15 h after line 12 fault

Under normal operating conditions, the methodology proposed in this chapter remains effective. This approach is applied to enhance the economic performance of the distribution system. Simulation results, presented in Fig. 9.10, indicate that the MES power sharing model, based on the P2P transaction framework, improves the operational efficiency of the distribution system. Compared to the conventional MES model, this approach enhances economic performance by enabling power sharing with other users through MES.

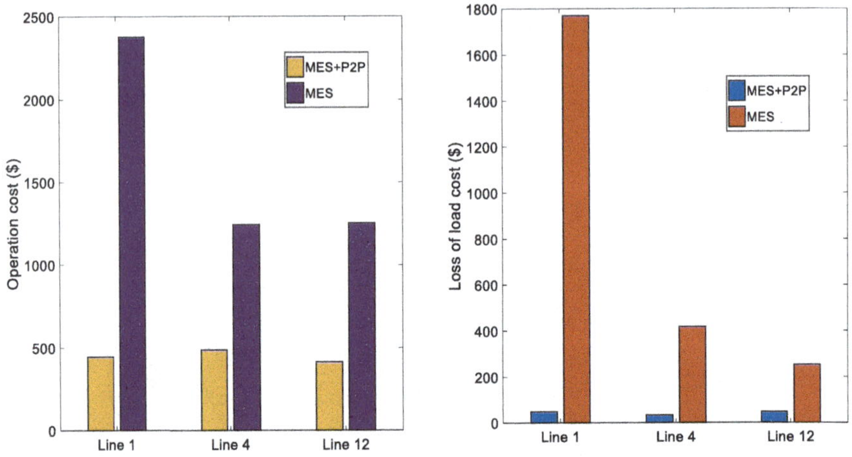

Fig. 9.8 Cost comparison of distribution system under different strategies

Table 9.6 Cost allocation of MSE-sharing based on proportional allocation method under each failure scenario

Faulty line	Line 1	Line 4	Line 12
Cost ($)	298.0	222.6	337.6
User 1	0.0	63.1	0.0
User 2	146.9	0.0	0.0
User 3	0.0	4.4	0.7
User 4	10.2	20.3	0.6
User 5	8.7	34.1	0.0
User 6	14.3	25.7	0.0
User 7	11.2	4.8	0.0
User 8	11.3	5.1	0.0
User 9	11.8	5.0	0.0
User 10	11.5	4.7	0.0
User 11	10.8	2.5	0.0
User 12	6.8	47.9	317.2
User 13	49.5	0.3	1.2
User 14	0.3	4.9	19.2
User 15	4.5	0.0	0.0

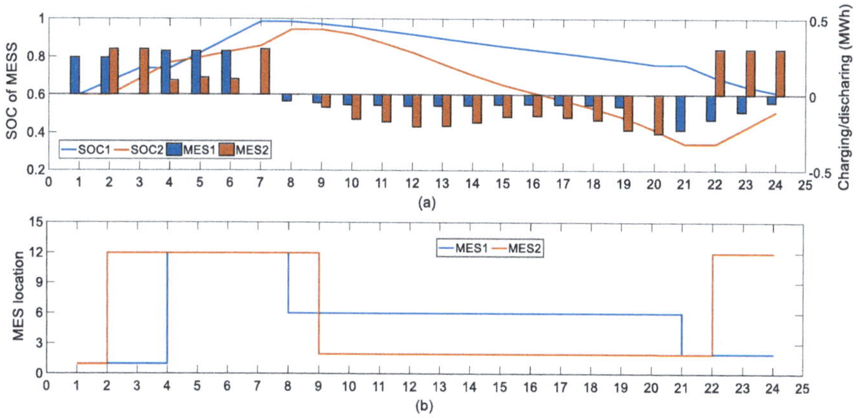

Fig. 9.9 Real-time state of MES after lines 1 and 4 fault

Table 9.7 Power supply when lines 1 and 4 fault

Load type	Load demand (MW)	Real load (MW)	Loss of load (MW)
Critical load	11.958	11.958	0.0
General load	3.984	3.224	0.740

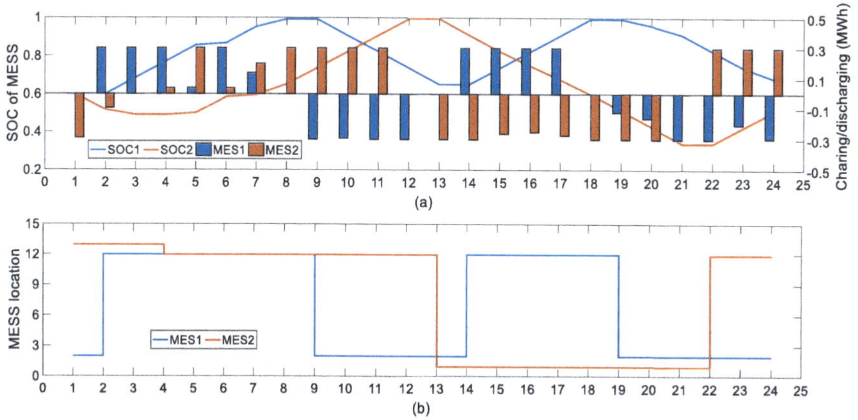

Fig. 9.10 Real-time state of MES under normal condition

9.6 Conclusion

This chapter proposes a resilience improvement strategy for distribution systems based on MES power sharing within a P2P transaction framework. The operational cost of the distribution system post-fault serves as the primary index for evaluating resilience. By leveraging the power sharing capabilities provided by the P2P transaction mode, the strategy aims to maximize the generalized benefits of MES, thereby enhancing the system's resilience.

A proportional distribution-based cost-sharing mechanism is introduced to determine the operation and maintenance costs each user owes to the MES owner. This mechanism calculates cost-sharing amounts based on the extent of power shared between users and MES, ensuring fair allocation when users participate in MES power sharing to varying degrees.

The proposed strategy is significant for improving the distribution system's ability to manage sudden faults and minimize fault-related losses. It provides a practical approach to enhancing system resilience through MES power sharing.

References

1. Nazemi M, Dehghanian P, Lu X et al (2021) Uncertainty-aware deployment of mobile energy storage systems for distribution grid resilience. IEEE Trans Smart Grid 12(4):3200–3214
2. Li J, Xu X, Yan Z et al (2023) Coordinated optimization of emergency response resources in transportation-power distribution networks under extreme events. IEEE Trans Smart Grid 14(6):4607–4620
3. Hou H, Tang J, Zhang Z et al (2023) Resilience enhancement of distribution network under typhoon disaster based on two-stage stochastic programming. Appl Energy 338:120892
4. Ding T (2016) Power system operation with large scale stochastic wind power integration: interval arithmetic based analysis and optimization methods. Springer
5. Shirazi E, Jadid S (2018) Autonomous self-healing in smart distribution grids using agent systems. IEEE Trans Industr Inf 15(12):6291–6301
6. Amini F, Ghassemzadeh S, Rostami N et al (2024) A stochastic two-stage microgrid formation strategy for enhancing distribution network resilience against earthquake event incorporating distributed energy resources, parking lots and responsive loads. Sustain Cities Soc 101:105191
7. Zhou Y, Li X, Han H et al (2024) Resilience-oriented planning of integrated electricity and heat systems: a stochastic distributionally robust optimization approach. Appl Energy 353:122053
8. Park JY, Ban J, Kim YJ et al (2021) Supplementary feedforward control of DGs in a reconfigurable microgrid for load restoration. IEEE Trans Smart Grid 12(6):4641–4654
9. Li B, Chen Y, Wei W et al (2021) Routing and scheduling of electric buses for resilient restoration of distribution system. IEEE Trans Transp Electrification 7(4):2414–2428
10. Paul S, Poudyal A, Poudel S et al (2024) Resilience assessment and planning in power distribution systems: past and future considerations. Renew Sustain Energy Rev 189:113991
11. Oboudi MH, Mohammadi M (2024) Two-stage seismic resilience enhancement of electrical distribution systems. Reliab Eng Syst Saf 241:109635
12. Nazemi M, Moeini-Aghtaie M, Fotuhi-Firuzabad M et al (2019) Energy storage planning for enhanced resilience of power distribution networks against earthquakes. IEEE Trans Sustain Energy 11(2):795–806
13. Lei S, Wang J, Chen C et al (2016) Mobile emergency generator pre-positioning and real-time allocation for resilient response to natural disasters. IEEE Trans Smart Grid 9(3):2030–2041
14. Huang Y, Jin M, Xie J et al (2024) Dynamic Bayesian game optimization method for multi-energy hub systems with incomplete load information. Energy 301:131622
15. Kim J, Dvorkin Y (2018) Enhancing distribution system resilience with mobile energy storage and microgrids. IEEE Trans Smart Grid 10(5):4996–5006
16. Baza M, Sherif A, Mahmoud MMEA et al (2021) Privacy-preserving blockchain-based energy trading schemes for electric vehicles. IEEE Trans Veh Technol 70(9):9369–9384

17. Chen W, Tanneau M, Van Hentenryck P (2023) End-to-end feasible optimization proxies for large-scale economic dispatch. IEEE Trans Power Syst
18. Guerrero J, Chapman AC, Verbič G (2018) Decentralized P2P energy trading under network constraints in a low-voltage network. IEEE Trans Smart Grid 10(5):5163–5173
19. Maneesha A, Swarup KS (2021) A survey on applications of alternating direction method of multipliers in smart power grids. Renew Sustain Energy Rev 152:111687
20. Morstyn T, Hredzak B, Agelidis VG (2016) Control strategies for microgrids with distributed energy storage systems: an overview. IEEE Trans Smart Grid 9(4):3652–3666
21. Kim J, Dvorkin Y (2019) A P2P-dominant distribution system architecture. IEEE Trans Power Syst 35(4):2716–2725
22. Sun W, Qiao Y, Liu W (2022) Economic scheduling of mobile energy storage in distribution networks based on equivalent reconfiguration method. Sustain Energy Grids Netw 32:100879
23. Ding T, Li C, Yang Y et al (2016) A two-stage robust optimization for centralized-optimal dispatch of photovoltaic inverters in active distribution networks. IEEE Transa Sustain Energy 8(2):744–754
24. Lotfi M, Osório GJ, Javadi MS et al (2021) A Dijkstra-inspired graph algorithm for fully autonomous tasking in industrial applications. IEEE Trans Ind Appl 57(5):5448–5460
25. Granelli G, Montagna M, Zanellini F et al (2006) A genetic algorithm-based procedure to optimize system topology against parallel flows. IEEE Trans Power Syst 21(1):333–340
26. Qu ZL, Chen JJ, Peng K et al (2021) Enhancing stochastic multi-microgrid operational flexibility with mobile energy storage system and power transaction. Sustain Cities Soc 71:102962
27. De Castro LN, Von Zuben FJ (2002) Learning and optimization using the clonal selection principle. IEEE Trans Evol Comput 6(3):239–251
28. Sotkiewicz PM, Vignolo JM (2006) Allocation of fixed costs in distribution networks with distributed generation. IEEE Trans Power Syst 21(2):639–652
29. Abdeltawab HH, Mohamed YARI (2017) Mobile energy storage scheduling and operation in active distribution systems. IEEE Trans Ind Electron 64(9):6828–6840
30. Lai K, Illindala MS (2021) Sizing and siting of distributed cloud energy storage systems for a shipboard power system. IEEE Trans Ind Appl 57(3):1935–1944
31. Liu Y, Wu X, Du J et al (2020) Optimal sizing of a wind-energy storage system considering battery life. Renew Energy 147:2470–2483
32. Hong Z, Wei Z, Li J et al (2021) A novel capacity demand analysis method of energy storage system for peak shaving based on data-driven. J Energy Storage 39:102617

The manufacturer's authorised representative in the EU is Springer Nature Customer Service Centre GmbH, Europaplatz 3, 69115 Heidelberg, Germany. If you have any concerns regarding our products, please contact ProductSafety@springernature.com

Printed and bound by CPI Group (UK) Ltd, Croydon, CR0 4YY
01/10/2026
02225922-0004